Fruit

pome
pear

achene
strawberry

berry
raspberry

drupe
peach

hesperidium
grapefruit

legume
runner bean

pepo
melon

seed pod
okra

nut
brazil

NICHOLAS LIGHT BSc PhD FIFST

LONGMAN ILLUSTRATED DICTIONARY OF FOOD SCIENCE

food, its components, nutrition,
preparation and preservation

LONGMAN ▦ ✿ YORK PRESS

YORK PRESS
Immeuble Esseily, Place Riad Solh, Beirut.

LONGMAN GROUP UK LIMITED
Burnt Mill, Harlow, Essex.

First published 1989

ISBN 0 582 02162 6

Illustrations by Charlotte Kennedy and Industrial Art Studio
Photocomposed in Britain by Panache, London
Printed and bound in Lebanon by Typopress, Beirut.

Contents

How to use the dictionary

This dictionary contains over 1200 words used in the food sciences.
These are arranged in groups under the main headings listed on pp. 3–4.
The entries are grouped according to the meaning of the words to help the
reader to obtain a broad understanding of the subject.

At the top of each page the subject is shown in bold type and the part of
the subject in lighter type. For example, on pp. 74 and 75:

74 · **VITAMINS**/VITAMINS E, K AND C

VITAMINS/VITAMIN B · **75**

In the definitions the words used have been limited so far as possible to
about 1500 words in common use. These words are those listed in the
'defining vocabulary' in the *New Method English Dictionary* (fifth edition) by
M. West and J.G. Endicott (Longman 1976). Words closely related to these
words are also used: for example, *characteristics*, defined under *character*
in West's *Dictionary*.

In addition to the entries in the text, the dictionary has several useful
appendixes which are detailed in the Contents list and are to be found at
the back of the dictionary.

---

1. To find the meaning of a word

Look for the word in the alphabetical index at the end of the book, then turn
to the page number listed.

In the index you may find words with a number at the end. These only
occur where the same word appears more than once in the dictionary in
different contexts. For example, **nucleus**

nucleus[1] is the centre of an atom;

nucleus[2] is a cell organelle.

The description of the word may contain some words with arrows in
brackets (parentheses) after them. This shows that the words with arrows
are defined near by.

(↑) means that the related word appears above or on the facing page;

(↓) means that the related word appears below or on the facing page.

A word with a page number in brackets after it is defined elsewhere in the
dictionary on the page indicated. Looking up the words referred to may
help in understanding the meaning of the word that is being defined.

In some cases more than one meaning is given for the same word. Where
this is so, the first definition given is the more (or most) common usage of
the word. The explanation of each word usually depends on knowing the
meaning of a word or words above it. For example, on p.92 the meaning of
Maillard browning, sugar amine and the words which follow depends on the
meaning of *non-enzymic browning,* which appears above them. Once the
earlier words are understood those that follow become easier to
understand. The illustrations have been designed to help the reader
understand the definitions but the definitions are not dependent on the
illustrations.

2. To find related words

Look in the index for the word you are starting from and turn to the page number shown. Because this dictionary is arranged by ideas, related words will be found in a set on that page or one near by. The illustrations will also help to show how words relate to one another.

For example, words relating to enzymes are on pp. 68–71. On p.68 *enzyme* is followed by *catalyst* and *catalysis* and there is an illustration showing the function of enzymes in the catalysis of reactions; p.69 continues to explain and illustrate enzymes explaining the different types of enzymes and illustrating the action of proteolytic enzymes; turning to p.70 inhibition is explained in its various forms and non-competitive inhibition is illustrated; and p.71 gathers together the remaining words relating to enzyme reactions.

3. As an aid to studying or revising

The dictionary can be used for studying or revising a topic. For example, to revise your knowledge of ingestion, you would look up *ingestion* in the alphabetical index. Turning to the page indicated, p.37, you would find *ingestion*, *mastication*, *saliva*, *mucin*, and so on; turning over to p.38 you would find *gastric juice*, *rennin*, *pepsin*, and so on. On p.39 you would find *jejunum*, *intestinal juice*, etc.

In this way, by starting with one word in a topic you can revise all the words that are important to this topic.

4. To find a word to fit a required meaning

It is almost impossible to find a word to fit a meaning in most dictionaries, but it is easy with this book. For example, if you had forgotten the word for the inner tissues that make up the pit of a fruit, all you would have to do would be to look up *pit* in the alphabetical index and turn to the page indicated, p.109. There you would find the word *endocarp* and a diagram to illustrate its meaning.

5. Abbreviations used in the definitions

abbr	abbreviation	p.	page
adj	adjective	pl	plural
e.g.	*exempli gratia* (for example)	pp	pages
etc	*et cetera* (and so on)	sing.	singular
i.e.	*id est* (that is to say)	v	verb
n	noun	=	the same as

THE
DICTIONARY

element (*n*) a natural substance containing atoms of the same kind. Elements are substances with their own characteristics and cannot be changed into other elements except by dividing or adding to the atoms of which they are made. Most elements can unite with other elements to form other substances by chemical reaction (p.12). The common elements found in food are carbon, hydrogen, oxygen and nitrogen.

symbol (*n*) the letter or letters which stand for the name of an element (↑), e.g. carbon (C), oxygen (O), hydrogen (H), nitrogen (N), chlorine (Cl).

formula (*n*) the set of letters and numbers which stand for the kind and number of atoms in a substance. Carbon dioxide, which contains one atom of carbon and two atoms of oxygen has the formula CO_2. **formulae** (*pl*).

atom (*n*) the smallest part of any element (↑). Atoms are made of protons (↓), electrons (↓) and neutrons (↓). Every atom contains equal numbers of electrons and protons. **atomic** (*adj*).

sub-atomic (*adj*) of particles (p.24) smaller than atoms.

symbol

C carbon

O oxygen

formula

CO_2 carbon dioxide

the four commonest atoms in biological compounds

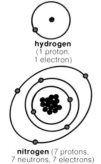

hydrogen
(1 proton,
1 electron)

nitrogen (7 protons,
7 neutrons, 7 electrons)

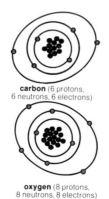

carbon (6 protons,
6 neutrons, 6 electrons)

oxygen (8 protons,
8 neutrons, 8 electrons)

neutron (*n*) a sub-atomic (↑) particle found in all atoms except hydrogen. The neutron has no electric charge and has the same weight as the proton (↓).

proton (*n*) a sub-atomic (↑) particle with a positive electric charge. The size of this electric charge is exactly equal to the size of the negative charge on the electron (↓). As atoms contain equal numbers of protons and electrons they have no charge.

electron (*n*) a sub-atomic (↑) particle 1840 times smaller than the proton (↑) which is in constant movement around the nucleus (↓) of the atom. The electron has a negative charge exactly equal to the positive charge of the proton. Atoms which lose or gain electrons become charged particles called ions (↓).

nucleus[1] (*n*) the central part of the atom which contains the protons (↑) and neutrons (↑).

orbital (*n*) the path that can be followed by each electron (↑) around the nucleus (↑) of the atom.

ion (*n*) an atom or molecule containing an unequal number of protons (↑) and electrons (↑) and, therefore, it has an electric charge. An ion with a positive charge has more protons than electrons and an ion with a negative charge has more electrons than protons. **ionization** (*n*), **ionizing** (*adj*), **ionic** (*adj*)

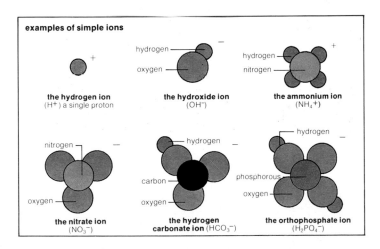

examples of simple ions

the hydrogen ion
(H^+) a single proton

hydrogen
oxygen
the hydroxide ion
(OH^-)

hydrogen
nitrogen
the ammonium ion
(NH_4^+)

nitrogen
oxygen
the nitrate ion
(NO_3^-)

hydrogen
carbon
oxygen
the hydrogen carbonate ion (HCO_3^-)

hydrogen
phosphorous
oxygen
the orthophosphate ion
($H_2PO_4^-$)

anion (*n*) an ion (p.9) with a positive charge.
cation (*n*) an ion (p.9) with a negative charge.

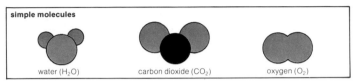

simple molecules

water (H_2O)   carbon dioxide (CO_2)   oxygen (O_2)

molecule (*n*) the smallest part of an element (p.8)
that can exist alone. Molecules contain atoms
held together by strong forces. A molecule of
water contains two hydrogen atoms and one
oxygen atom (H_2O). The smallest molecule is
hydrogen which contains two hydrogen atoms.
molecular (*adj*).

macromolecule (*n*) any large molecule containing
many atoms, e.g. nucleic acids (p.31),
carbohydrates (p.43), proteins (p.61).

biological macromolecules

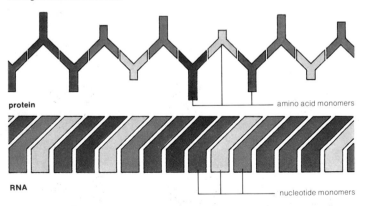

protein   amino acid monomers

RNA   nucleotide monomers

polymer (*n*) a macromolecule (↑) containing many
molecules of the same sort joined together, e.g.
starch (p.48), cellulose (p.49). Polypeptides
(p.63) and nucleic acids (p.31) are sometimes
called polymers but are more properly called
macromolecules.

monomer (*n*) a single molecule from which a polymer (↑) is made, e.g. monosaccharides (p.43), amino acids (p.61), nucleotides (p.31).

crystal (*n*) a solid structure with a regular arrangement of molecules all of the same kind and size. **crystalline** (*adj*), **crystallize** (*v*).

compound (*n*) a substance which contains one kind of molecule consisting of more than one atom, e.g. common salt (NaCl).

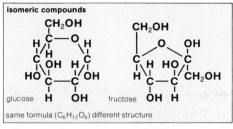

isomeric compounds

glucose

fructose

same formula ($C_6H_{12}O_6$) different structure

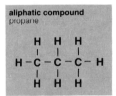

aliphatic compound
propane

aromatic compound
benzoic acid

mixture (*n*) a substance which contains more than one kind of molecule, e.g. air is a mixture of carbon dioxide (CO_2), oxygen (O_2) and nitrogen (N_2).

isomer (*n*) molecules which have the same kinds and numbers of atoms. Isomers may have different chemical and physical properties. **isomeric** (*adj*).

organic (*adj*) of compounds built around the carbon atom. Not all compounds containing carbon are organic, e.g. carbonates ($CaCO_3$, calcium carbonate). All living organisms are made mainly of water and organic compounds.

inorganic (*adj*) of compounds which are not organic (↑).

aliphatic (*adj*) of organic (↑) compounds which contain chains of carbon atoms.

aromatic (*adj*) of organic (↑) compounds which contain carbon atoms joined together in rings. Aromatic compounds often have a characteristic smell.

hydrocarbon (*n*) an organic (↑) compound containing only carbon and hydrogen atoms.

alkane (*n*) a hydrocarbon (p.11) containing only saturated (p.18) or single bonds (p.18). Many of the simple alkanes are used as fuels, e.g. methane, propane, butane.

alkene (*n*) a hydrocarbon (p.11) containing double bonds (p.18) between carbon atoms. Ethene (p.110), a gas made by plants which speeds the ripening of fruits, is an alkene.

alkyne (*n*) a hydrocarbon (p.11) containing triple bonds (p.18) between carbon atoms. The simplest alkyne is the gas ethyne (acetylene) which is used as a fuel.

reaction

$$A + B \rightleftharpoons C + D$$

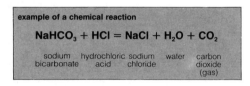

example of a chemical reaction

$$NaHCO_3 + HCl = NaCl + H_2O + CO_2$$

sodium   hydrochloric sodium   water   carbon
bicarbonate acid   chloride   dioxide
                                                 (gas)

reaction (*n*) a chemical change in which one or more molecules produce a different molecule or molecules. Reactions always either use or give off energy. The molecules in a reaction may be the same or different kinds. **reactive** (*adj*), **react** (*v*).

exergonic (*adj*) of a chemical reaction (↑) which gives off energy, e.g. as heat.

endergonic (*adj*) of a chemical reaction (↑) which needs energy to drive it, e.g. the reacting chemicals become cold as they take up heat from their surroundings.

equilibrium (*n*) the point at which a chemical reaction (↑) stops.

inert (*adj*) of a substance or element (p.8) which does not react (↑) easily with any other substances or elements, e.g. the noble gases such as helium (He) and argon (Ar). Some substances used in the food industry are inert, e.g. PTFE (p.164) which is used in the production of plastics (p.165) for wrapping food.

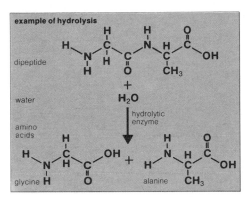

example of hydrolysis

dipeptide

water

hydrolytic enzyme

amino acids

glycine

alanine

hydrolysis (*n*) a chemical reaction (↑) where a substance is broken down and water is added to one or more of its atoms or molecules. **hydrolyse** (*v*), **hydrolytic** (*adj*).

synthesis (*n*) a chemical reaction (↑) where molecules are built into larger molecules and macromolecules (p.10) e.g. proteins (p.61) from amino acids (p.61). **synthesize** (*v*), **synthetic** (*adj*).

condensation reaction a reaction (↑) where two molecules are joined together with the loss of one molecule of water.

example of a condensation reaction

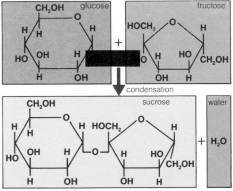

glucose

fructose

condensation

sucrose

water

reduction[1] (*n*) a reaction (p.12) in which a molecule (1) gains electrons (p.9); (2) loses oxygen; or (3) gains hydrogen. **reduce** (*v*), **reductive** (*adj*).

oxidation (*n*) a reaction (p.12) in which a molecule (1) loses electrons (p.9); (2) loses hydrogen; or (3) gains oxygen. **oxidize** (*v*), **oxidative** (*adj*).

redox (*adj*) of chemical reactions (p.12) where reduction (↑) and oxidation (↑) take place.

potential energy energy which is stored in molecules and can be given off during a chemical reaction (p.12).

acid (*n*) a compound which can lose protons (p.9), i.e. hydrogen (H^+) ions (p.9), and donate (p.16) them to water forming H_3O^+ ions. **acidic** (*adj*), **acidify** (*v*), **acidity** (*n*).

pH

$$pH = \log_{10} \frac{1}{[H^+]} \text{ or } - \log_{10} [H^+]$$

where [**H**$^+$] = hydrogen ion concentration

pH scale

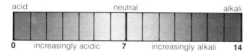

acid   neutral   alkali

0   increasingly acidic   **7**   increasingly alkali   **14**

pH the negative logarithm of hydrogen ion (H^+) concentration (p.22). The scale of measurement of pH ranges from 0 for a hydrogen ion, concentration of 1M to 14 for a hydrogen ion concentration of 10^{-14}M. The pH scale is used to measure the acidity (pH 0 to 7) or alkalinity (pH 7 to 14) of compounds or mixtures.

indicator (*n*) a compound which shows the acidity (↑) or alkalinity (↓) of a substance by changing colour.

strong acid an acid (↑) whose ions (p.9) fully dissociate (↓) in solution (p.22), e.g. hydrochloric acid (HCl), nitric acid (HNO_3).

weak acid an acid (↑) whose ions (p.9) only partly dissociate (↓) in solution (p.22), e.g. ethanoic acid (CH_3COOH).

dissociation

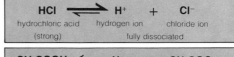

HCl ⇌ H⁺ + Cl⁻
hydrochloric acid   hydrogen ion   chloride ion
(strong)   fully dissociated

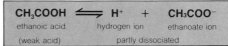

CH₃COOH ⇌ H⁺ + CH₃COO⁻
ethanoic acid   hydrogen ion   ethanoate ion
(weak acid)   partly dissociated

dissociation (*n*) the way in which ionic (p.9)
compounds in solution (p.22) split into the ions
(p.9) from which they are made. For example,
when common salt, (NaCl) is dissolved (p.22) it
dissociates to form free sodium ions (Na⁺) and
free chloride ions (Cl⁻). **dissociate** (*v*).

base (*n*) a compound which can accept protons
(p.9), i.e hydrogen (H⁺) ions (p.9). **basic** (*adj*).

alkali (*n*) a base (↑). Alkalis, like sodium hydroxide
(NaOH), dissociate (↑) to form hydroxyl (OH⁻)
ions (p.9).

strong base a base (↑) whose ions (p.9) fully
dissociate (↑) in solution (p.22), e.g. sodium
hydroxide (NaOH).

weak base a base (↑) whose ions (p.9) only partly
dissociate (↑) in solution (p.22), e.g. disodium
orthophosphate (Na₂HPO₄).

Bronsted acid a proton (p.9) donor (p.16). Acid-
base reactions (p.12) always involve a conjugate
acid-base pair (↓) made up of a proton donor
(acid (↑)) and a proton acceptor (p.16) (base (↑)).

conjugate acid-base pair the mixture of a proton
(p.9) donor (p.16) and proton acceptor (p.16)
which is always involved in an acid-base reaction
(p.12). For example, ethanoic acid (CH₃COOH)
and the ethanoate ion (CH₃COO⁻) form a
conjugate acid-base pair.

conjugate acid-base pair

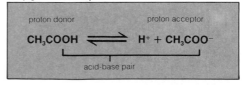

proton donor   proton acceptor

CH₃COOH ⇌ H⁺ + CH₃COO⁻

acid-base pair

neutral (*adj*) of a solution (p.22) which is neither acidic (p.14) nor basic (p.15). Neutral solutions have a pH (p.14) of 7.0. **neutralize** (*v*).

titrate (*v*) to add a base (p.15) to an acid (p.14) or an acid to a base until the required pH (p.14) is reached. The word can also be used of any reaction (p.12) where one compound is added to another or a mixture of others until an end point is reached. **titration** (*n*).

donor (*n*) a molecule which can give up part of itself to another molecule. **donate** (*v*).

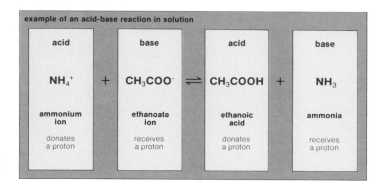

example of an acid-base reaction in solution

acid	base	acid	base
NH_4^+	CH_3COO^- $\rightleftharpoons$	CH_3COOH	NH_3
ammonium ion	ethanoate ion	ethanoic acid	ammonia
donates a proton	receives a proton	donates a proton	receives a proton

acceptor (*n*) a molecule which can accept a part of a different molecule.

salt[1] (*n*) an ionic (p.9) compound produced by the reaction (p.12) of an acid (p.14) with a base (p.15) or a metal with an acid.

mineral elements elements (p.8) other than carbon, hydrogen, oxygen, nitrogen, phosphorus and sulphur which are needed by living organisms.

minerals (*n.pl.*) = mineral elements (↑).

trace element any element (p.8) found in or required by living organisms in very low quantities, e.g. molybdenum (Mb).

interaction (*n*) the way in which actual compounds act on each other without chemical reaction (p.12), e.g. proteins (p.61) can interact with water and hold on to it very strongly. **interact** (*v*).

bond (*n*) a link between atoms or molecules.
Bonds can be of various kinds, e.g. covalent (↓),
non-covalent (↓), ionic (↓) and hydrophobic
(p.18).

covalent bond a chemical bond (↑) between atoms
in which electrons (p.9) in outer orbitals (p.9) are
shared.

ionic bond a strong bond (↑) in which electrons
(p.9) are donated (↑) by one atom to a different
atom, e.g. in sodium chloride (NaCl) the sodium
atom donates one electron to the chloride atom
to form a stable compound.

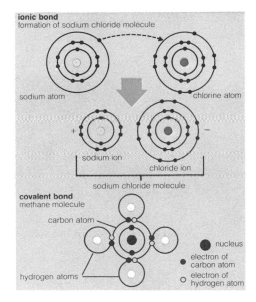

ionic bond
formation of sodium chloride molecule

sodium atom · chlorine atom

+ sodium ion −
chloride ion

sodium chloride molecule

covalent bond
methane molecule

carbon atom

nucleus
electron of
carbon atom
electron of
hydrogen atom

hydrogen atoms

non-ionic (*adj*) of a chemical bond (↑) formed
between atoms or molecules in which
electrons (p.9) are neither donated (↑) nor
accepted (↑).

non-covalent bond any chemical bond (↑) which
is not covalent (↑), e.g. hydrogen bond (p.18),
electrostatic bond (p.18), hydrophobic bond
(p.18).

hydrogen bond a weak non-covalent bond (p.17) formed between a hydrogen atom in one molecule and an atom of oxygen or nitrogen in a different molecule.

electrostatic bond a non-covalent bond (p.17) formed between ionic (p.9) atoms of different electrical charge.

hydrophobic bond a non-covalent bond (p.17) formed between molecules which cannot mix with water. Such molecules are oily (p.52) or fatty (p.52) in nature and interact (p.16) with each other to keep water out.

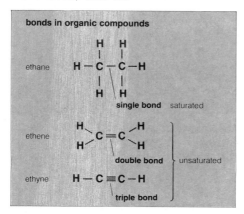

bonds in organic compounds

ethane — single bond — saturated

ethene — double bond — unsaturated

ethyne — triple bond

single bond a covalent bond (p.17) in which only one pair of electrons (p.9) are shared between two atoms.

double bond a covalent bond (p.17) in which two pairs of electrons (p.9) are shared between two atoms.

triple bond a covalent bond (p.17) in which three pairs of electrons (p.9) are shared between two atoms.

saturated bond a single bond (p.17).

unsaturated bond a double (↑) or triple bond (↑). In organic (p.11) compounds unsaturated bonds are usually formed between two atoms of carbon ($-C=C-$), carbon and nitrogen atoms ($-C=N-$) or carbon and oxygen atoms ($-C=O$).

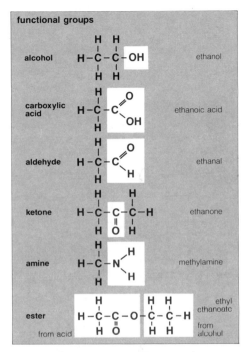

functional groups

alcohol — H–C–C–OH — ethanol

carboxylic acid — H–C–C(=O)OH — ethanoic acid

aldehyde — H–C–C(=O)H — ethanal

ketone — H–C–C(=O)–C–H — ethanone

amine — H–C–N(H)(H) — methylamine

ester — H–C–C(=O)–O–C–C–H — ethyl ethanoate
from acid ··· from alcohol

functional group the part of an organic (p.11) molecule which gives that compound its main properties. There are five important ones in food compounds: (1) hydroxyl or alcohol (–OH); (2) amine (–NH$_2$); (3) aldehyde (–CHO); (4) ketone (–CO); (5) carboxylic acid (–COOH).

ester (*n*) a compound formed when a carboxylic acid (–COOH) reacts (p.12) with an alcohol (–OH). Ethyl ethanoate (also known as ethyl acetate, CH$_3$COOC$_2$H$_5$) is an example of an ester. Many food macromolecules (p.10) contain or form ester bonds (p.17), e.g. fats (p.52), proteins (p.61), carbohydrates (p.43).

derivative (*n*) a molecule produced by adding one or more functional groups (↑) by chemical reaction (p.12) to another molecule.

chemical analysis the process by which the chemical content and structure of a substance is discovered. Chemical analysis is done by a series of chemical reactions (p.12) and tests which are used to determine the chemical content and structure of the substance.

ash (*n*) the part of a food which remains after all the organic (p.11) substances have been burnt off. Ash is used as a measure of the amount of inorganic (p.11) compounds in the food.

halogen (*n*) the common name for the group of elements (p.8), fluorine (F), chlorine (Cl), bromine (Br) and iodine (I). Halogens form compounds with metals, e.g. sodium chloride (NaCl) and organic (p.11) compounds. Chlorine is the most common halogen in foods but iodine, bromine and fluorine are also found in small amounts.

phosphate (*n*) an inorganic (p.11) ion (p.9) present in all living organisms. Phosphate (PO_4^{3-}) is found in ATP (p.36) and the nucleotide (p.31) molecules of nucleic acids (p.31).

sulphate (*n*) an inorganic (p.11) ion (p.9) important in the diet (p.137). Sulphate (SO_4^{2-}) is added as a derivative (p.19) to some macromolecules (p.10) such as polysaccharides (p.44).

nitrate (*n*) an important inorganic (p.11) ion (p.9) found in the soil and used by plants as a nutrient (p.137). Nitrate (NO_3^-) as well as nitrite (NO_2^-) is used in the food industry as a preservative (p.155).

nitrogen cycle the cycle of nitrogen through nature. Amino acids (p.61) and proteins (p.61) as well as many other organic (p.11) molecules contain nitrogen. Plants take up nitrogen from the soil through their roots in the form of nitrates (↑) and use it to make proteins and other molecules. Animals eat plants, use the plant nitrogen for their own proteins and other organic molecules and excrete waste nitrogen. Nitrogen also returns to the soil when animals and plants die and decompose.

saltpetre (*n*) the common name for potassium nitrate (KNO_3) used as a meat preservative (p.155).

Chile saltpetre the common name for sodium nitrate ($NaNO_3$) used as a meat preservative (p.155).

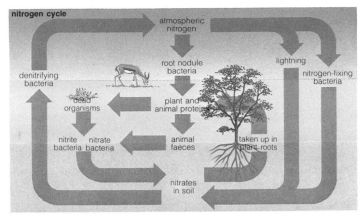

ethanoic acid a weak acid (p.14) found in vinegar. Ethanoic acid (CH_3COOH) is formed from ethanol (↓) by certain bacteria (p.148). Also known as **acetic acid.**

citric acid the main acid of fruit, particularly citrus fruit (p.108). Citric acid ($CH_2COOHCOHCOOHCH_2COOH$) is also an important compound in metabolism (p.144).

lactic acid (n) a weak acid (p.14) found in muscle (p.95). It is produced when muscles work without a good supply of oxygen, i.e. under anaerobic (p.32) conditions. Lactic acid ($CH_3CHOHCOOH$) is also formed when milk sours by the action of bacteria (p.148) on lactose (p.47)

malic acid $HOOCCH_2CHOHCOOH$. A weak acid (p.14) found in fruits, e.g. apples and pears.

tartaric acid $HOOCCHOHCHOHCOOH$. A weak acid (p.14) found in fruit especially grapes. Tartaric acid and its salt (p.16) potassium hydrogen tartrate (*cream of tartar*) are used in baking powders (p.121) as the acid (p.14).

ethanol (n) C_2H_5OH. The most common food alcohol formed by the action of yeast (p.151) on sugar (p.44) in the process of fermentation (p.34). Ethanol is the main alcohol in all alcoholic beverages (p.93) such as beer (p.93), cider (p.93), wine (p.93). Also known as **ethyl alcohol.**

gas (*n*) an air-like substance which is neither solid (↓) nor liquid (↓) at normal temperatures. **gaseous** *(adj)*.

liquid (*n*) a substance which is neither gas (↑) nor solid (↓) at normal temperatures. Liquids have no form and can flow. **liquefy** (*v*).

solid (*n*) a substance which is neither gas (↑) nor liquid (↑) at normal temperatures. Solids are firm and can keep their shape. **solidify** (*v*).

vacuum (*n*) a place or region containing no solid (↑), liquid (↑) or gas (↑).

partial vacuum space from which most of the gas (↑) has been removed.

phase (*n*) the state of a substance, e.g. water can be fluid (liquid (↑) phase), ice (solid (↑) phase) or steam (gaseous (↑) phase).

dissolve (*v*) to spread the molecules of a solid (↑) or gas (↑) throughout a liquid (↑). Substances which dissolve in liquids are generally made of small molecules. In most cases, liquids are said to mix with other liquids, not to dissolve in them.

solution (*n*) a liquid (↑) which contains one or more dissolved (↑) substances.

solvent (*n*) a liquid (↑) in which molecules are dissolved (↑) to form a solution (↑).

solute (*n*) the molecules which are dissolved (↑) to form a solution (↑).

concentration (*n*) a measure of the number of solute (↑) molecules in a solution (↑) or the number of gas (↑) molecules in a given volume. Concentration can be expressed in units of grams per litre (gl^{-1}) but is normally expressed in units of molarity (↓).

molarity (*n*) a unit of concentration (↑). A one molar (1M) solution (↑) of a compound contains the molecular weight in grams of that compound dissolved (↑) in one litre of solution. For example, a 0.25M solution of NaCl (molecular weight 58.55) contains 14.66 gl^{-1} of dissolved NaCl.

concentrate (*v*) to increase the number of solute (↑) molecules in a solution (↑) or the number of gas (↑) molecules in a given volume. *See also* concentration (↑).

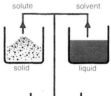

solution

solute   solvent

solid   liquid

solution

liquid

viscous (*adj*) of a solution (↑) or liquid (↑) which is thick and hard to pour. In foods, sauces (p.90) and syrups (p.50) are viscous. Foods can be made more viscous by adding thickeners (p.84) *See also* viscosity (p.27).

dilute (*v*) to decrease the concentration (↑) of a solute (↑) in a solution (↑) by adding more solvent (↑) in a given volume or by removing solute molecules from a given volume. **dilute** (*adj*).

soluble (*adj*) of a substance which is able to be dissolved (↑).

insoluble (*adj*)of a substance which is not able to be dissolved (↑).

aqueous (*adj*)of a solution (↑) where the solvent (↑) is water.

dispersion (*adj*) a gas (↑), liquid (↑) or solid (↑) which is made up of molecules, aggregates (p.25) of molecules or particles (p.24) of one kind spread among molecules of another kind. Dispersions can consist of solids in liquids, solids in gases, liquids in liquids, gases in liquids, gases in solids and liquids in solids. Solutions (↑), suspensions (↓) and colloids (p.25) can all be called dispersions. **disperse** (*v*)

suspension (*n*) dispersion (↑) of particles (p.24) in a liquid (↑) or a solution (↑). If left standing the particles in a suspension will settle to the bottom of the container. The particles in a suspension can also be filtered (p.24) out.

differences between solutions and suspensions

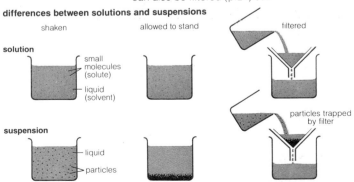

shaken · allowed to stand · filtered

solution — small molecules (solute), liquid (solvent)

suspension — liquid, particles, particles trapped by filter

filter (v) to separate particles (↓) in a suspension (p.23) from the liquid (p.22) in which they are dispersed (p.23) by passage through a semi-permeable (p.28) membrane (p.31).

particle (n) a small solid (p.22) piece of a substance, e.g. grains of sand, starch granule (p.49), bacteria (p.148), viruses (p.147). When particles are dispersed (p.23) in a liquid (p.22) they form suspensions (p.23) or colloids (↓).

homogeneous (adj) of solids (p.22), liquids (p.22) or gases (p.22) which are the same throughout, i.e. have an even kind and dispersion (p.23) of molecules. Solutions (p.22) are homogeneous. **homogeneity** (n).

heterogeneous (adj) of solids (p.22), liquids (p.22) or gases (p.22) which are uneven or mixed. Suspensions (p.23) are heterogeneous. **heterogeneity** (n).

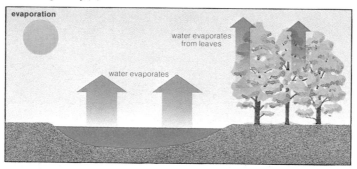

evaporation

water evaporates from leaves

water evaporates

evaporation (n) the loss of molecules from a liquid (p.22) or solution (p.22) as vapour (↓). The process by which molecules pass from the solid (p.22) or liquid (p.22) phase (p.22) to the gaseous phase. Steam rising from heated water is an example of evaporation.

vapour (n) the gaseous (p.22) form of molecules which are solids (p.22) or liquids (p.22) at normal temperature, e.g. steam is water vapour.

volatile (adj) of molecules which evaporate (↑) easily.

condensation (n) the formation of a liquid (p.22) or solid (p.22) from its vapour (↑).

some important food colloids

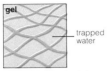

macromolecules (proteins, polysaccharides)

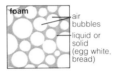

large macromolecules (egg white proteins)

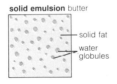

solid emulsion butter

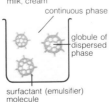

liquid emulsion
milk, cream

colligative properties the physical properties of solutions (p.22) which depend on the concentration (p.22) of solute (p.22) molecules, i.e. vapour pressure ($\downarrow$), boiling point ($\downarrow$), freezing point ($\downarrow$).

vapour pressure the pressure of a gas on a liquid (p.22) when the rate of evaporation ($\uparrow$) and condensation ($\uparrow$) are equal. Vapour pressure increases at higher temperatures but it decreases with higher solute (p.22) concentrations (p.22).

boiling point the temperature at which vapour pressure ($\uparrow$) is just higher than outside pressure so that a liquid (p.22) quickly evaporates ($\uparrow$), i.e. boils. The boiling point becomes higher with increasing solute (p.22) concentration (p.22).

freezing point the temperature at which a solution (p.22) or liquid (p.22) solidifies (p.22), i.e. freezes. Freezing point becomes lower with increasing solute concentration (p.22).

colloid (n) a dispersion (p.23) of macromolecules (p.10), aggregates ($\downarrow$) of macromolecules, gas (p.22) bubbles or particles ($\uparrow$) in a liquid (p.22), solid (p.22) or gas. The particles in a colloidal dispersion are aggregates of molecules and have sizes of 1–100 nm. A colloid is not a suspension (p.23) as the particles in a colloid will not settle out if left to stand and cannot be filtered ($\uparrow$).

aggregate (n) a gathering of molecules, particles ($\uparrow$) or cells. **aggregate** (v), **aggregation** (n).

continuous phase the liquid (p.22), solid (p.22) or gas (p.22) phase (p.22) in which molecular aggregates ($\uparrow$) are dispersed (p.23) to form a colloid ($\uparrow$).

disperse phase the molecular aggregates ($\uparrow$) or gas (p.22) bubbles which are dispersed (p.23) in the continuous phase (p.22) to form a colloid ($\uparrow$).

gel (n) a colloid ($\uparrow$) in which the disperse phase ($\uparrow$) is a liquid (p.22) and the continuous phase ($\uparrow$) is a solid (p.22), e.g. jelly.

gelation (n) the process by which a gel ($\uparrow$) is formed. It normally involves the trapping of water within a matrix of macromolecules (p.10), e.g. gelatin (p.98), agar (p.85), starch (p.48).

solid foam a colloid (p.25) in which the disperse phase (p.25) is a gas (p.22) and the continuous phase (p.25) is a solid (p.22), e.g. meringue.

sol (*n*) a colloid (p.25) in which the disperse phase (p.25) is a solid (p.22) and the continuous phase (p.25) is a liquid (p.22), e.g. proteins (p.61) in aqueous (p.23) solutions (p.22). Egg white and the cytoplasm (p.29) of cells are examples of sols.

solid sol a colloid (p.25) in which the disperse phase (p.25) and the continuous phase (p.25) are both solids (p.22), e.g. candy.

emulsion (*n*) a colloid (p.25) in which the disperse phase (p.25) is a liquid (p.22) and the continuous phase (p.25) is another, immiscible (↓) liquid. The immiscible liquids in most emulsions are hydrophobic (↓) and hydrophilic (↓) liquids. Emulsions are very common in natural and processed foods. They can be dispersions (p.23) of oil (p.52) in water or an aqueous (p.23) solution (p.22), (e.g. French dressing, mayonnaise), or dispersions of water or an aqueous solution in oil or fat (e.g. butter, margarine). Many emulsions are unstable (p.167) and need special molecules called emulsifiers (p.84) or surfactants (p.84) to make them stable (p.167).

foam (*n*) a colloid (p.25) in which the disperse phase (p.25) is a gas (p.22) and the continuous phase (p.25) is a liquid (p.22), e.g. the 'head' on beer, whipped egg white.

aerosol (*n*) a colloid (p.25) in which the disperse phase (p.25) is a solid (p.22) or a liquid (p.22) and the continuous phase (p.25) is a gas (p.22), e.g. fog, mist.

hydrophobic and hydrophilic surfactant molecules

hydrophobic tail

hydrophilic head

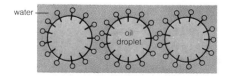

water

oil droplet

immiscible (*adj*) of two or more liquids (p.22) which will not mix.

hydrophobic (*adj*) of any substance which repels or will not mix with water.

hydrophilic (*adj*) of any substance which easily combines or mixes with water.

lyophobic (*adj*) of a colloid (p.25) in which the particles (p.24) of the disperse phase (p.25), do not mix easily with the continuous phase (p.25), e.g. emulsions (↑).

lyophilic (*adj*) of a colloid (p.25) in which the particles (p.24) of the disperse phase (p.25) mix easily with the continuous phase (p.25), e.g. sols (↑), gels (p.25).

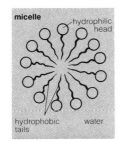

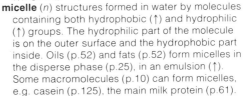

micelle / hydrophilic head / hydrophobic tails / water

micelle (*n*) structures formed in water by molecules containing both hydrophobic (↑) and hydrophilic (↑) groups. The hydrophilic part of the molecule is on the outer surface and the hydrophobic part inside. Oils (p.52) and fats (p.52) form micelles in the disperse phase (p.25), in an emulsion (↑). Some macromolecules (p.10) can form micelles, e.g. casein (p.125), the main milk protein (p.61).

texture (*n*) the structure of a food which accounts for its feel or consistency.

viscosity (*n*) a measure in liquids (p.22) of their ability to flow. Viscous (p.23) liquids are ones which are thick and do not flow easily.

water activity a measure of the availability of water in a food for microbial (p.147) growth or chemical reaction (p.12). Water activity (a_w) is the ratio of the vapour pressure (p.25) of water in the food to that of pure water at the same temperature. It is usually measured on a scale from 0 to 1.0. Foods which have been dried (e.g. raisins, sultanas, dried milk), have low water activities (0.3 – 0.6) and can be preserved (p.155) for a long time.

free water the water in a food which can easily be taken out by evaporation (p.24) or dehydration (p.162).

bound water the water in a food which is bonded (p.17) either strongly or weakly to other compounds but is not free water (↑). Bound water cannot act as a solvent (p.22) for other compounds in the food and is not easily taken out by dehydration (p.162).

syneresis (*n*) the shrinkage of a gel (p.25) causing the loss of liquid (p.22) from it.

osmosis (*n*) the passing of solvent (p.22) molecules through a semi-permeable (↓) membrane (p.31) from a solution (p.22) of low solute (p.22) concentration (p.22) to one of higher solute concentration.

osmosis

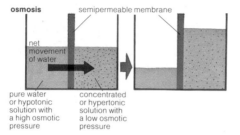

semipermeable membrane

net movement of water

pure water or hypotonic solution with a high osmotic pressure

concentrated or hypertonic solution with a low osmotic pressure

semi-permeable (*adj*)of a membrane (p.31) which can allow solvent (p.22) molecules (usually water) to pass through but not larger molecules, e.g. proteins (p.61) polysaccharides (p.44) dissolved in the solvent.

diffusion (*n*) the movement of molecules of a solute (p.22) from areas of higher concentration (p.22) to those of lower concentration.

hydration (*n*) the adding of water to any dry substance. **hydrate** (*v*).

re-hydration (*n*) the adding of water to any dried food or substance to return it to its normal state. **re-hydrate** (*v*).

humidity (*n*) the moisture content of the air or of the atmosphere in a container.

relative humidity a way of measuring the water vapour (p.24) in the air. It is the amount of water vapour in a given volume of air relative to the amount of water vapour in the same volume of air when it is saturated with water vapour (at the same temperature). Relative humidity (RH) is expressed as a percentage and can be shown mathematically as, $RH = M/M_s \times 100$, where M=mass of water in a volume of air, M_s=mass of water in the same volume of saturated air at the same temperature.

dehydration

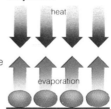

heat

evaporation

grapes

dehydrated grapes now called sultanas or raisins

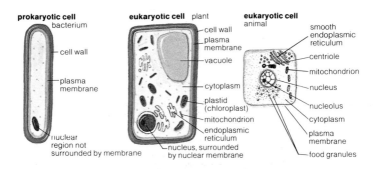

prokaryotic cell
 bacterium
  cell wall
  plasma membrane
 nuclear region not surrounded by membrane

eukaryotic cell   plant
  cell wall
  plasma membrane
  vacuole
  cytoplasm
  plastid (chloroplast)
  mitochondrion
  endoplasmic reticulum
  nucleus, surrounded by nuclear membrane

eukaryotic cell   animal
  smooth endoplasmic reticulum
  centriole
  mitochondrion
  nucleus
  nucleolus
  cytoplasm
  plasma membrane
  food granules

cell (*n*) a minute bag-like structure surrounded by a membrane (p.31) and containing DNA (p.31) and small bodies called organelles (↓). Apart from viruses (p.147) all living things are made up of one or more cells. Plant cells have a cell wall (p.32) as well as a membrane which surrounds the whole cell. There are two kinds of cells, eukaryotic (↓) and prokaryotic (↓). **cellular** (*adj*).

prokaryotic (*adj*) of cells having no membrane (p.31) surrounding the nuclear (p.30) material and no organelles (↓), e.g. bacteria (p.148). **prokaryote** (*n*).

eukaryotic (*adj*) of cells containing organelles (↓) and having a nucleus (p.30). The cells of all higher plants and animals are eukaryotic. **eukaryote** (*n*).

cytoplasm (*n*) that part of the cell which is inside the membrane (p.31), but is outside the nucleus (p.30). The cytoplasm is a fluid in the form of a sol (p.26).

intracellular (*adj*) of the inside of a cell.

extracellular (*adj*) of the outside of a cell.

organelle (*n*) a small body inside a eukaryotic (↑) cell. Organelles are normally contained within membranes (p.31) and have special functions, e.g. chloroplasts (p.30), mitochondria (p.30).

vacuole (*n*) a space in a cell which is filled with liquid (p.22) and surrounded by a membrane (p.31). Larger vacuoles are typical of plant cells.

vesicle (*n*) a small body which is surrounded by a
   membrane (↓) and found in a cell or organelle
   (p.29) usually containing the products of
   metabolism (p.144).

mitochondrion (*n*) a rod-shaped organelle (p.29)
   where the reactions (p.12) of the Krebs cycle
   (p.34) and the electron transfer chain (p.34) take
   place. Mitochondria have a smooth outer
   membrane (↓) and a folded inner membrane.
   mitochondria *(pl)*.

plastid (*n*) a kind of organelle (p.29) found in plant
   cells which is surrounded by a double membrane
   (↓). There are several kinds of plastid, each with
   a different function, e.g. chloroplasts (↓).

chloroplast

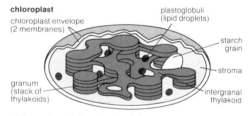

chloroplast envelope (2 membranes)
plastoglobuli (lipid droplets)
starch grain
stroma
granum (stack of thylakoids)
intergranal thylakoid

chloroplast (*n*) a plastid (↑) which contains the
   green pigment (p.78) chlorophyll (↓) and is
   where some stages of photosynthesis (p.43) are
   carried out. Chloroplasts contain their own DNA
   (↓) and can reproduce (p.107) themselves.

chlorophyll (*n*) the green-coloured pigment (p.78)
   found in plants which traps light energy to be
   used in photosynthesis (p.43). It is found in
   chloroplasts (↑) in the cells of leaf tissue (p.32)
   and green stems (p.104).

chlorophyllase (*n*) an enzyme (p.68) found in all
   plants which hydrolyses (p.13) chlorophyll (↑).

chromoplast (*n*) a plastid (↑) which contains
   pigments (p.78), e.g. in fruits.

nucleus[2] (*n*) an organelle (p.29) of eukaryotic
   (p.29) cells which contains the DNA (↓) and
   smaller organelles called nucleoli. Most cells
   contain only one nucleus. **nuclear** (*adj*).

multinucleate (*adj*) of cells containing more than
   one nucleus (↑), e.g. skeletal muscle (p.95)
   cells, the hyphae (p.151) of fungi (p.150).

nucleus

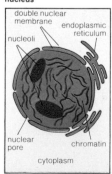

double nuclear membrane
endoplasmic reticulum
nucleoli
nuclear pore
chromatin
cytoplasm

nucleolus (*n*) a small vesicle (↑) containing RNA
(↓) found in the nucleus (↑). **nucleoli** (*pl*).

nucleic acid a macromolecule (p.10) of repeating
nucleotide (↓) units in long chain form. There are
two kinds of nucleic acid, DNA (↓) and RNA (↓).
These are used in the cell for the synthesis (p.13)
of proteins (p.61).

nucleotide (*n*) the monomers (p.11) from which
nucleic acids (↑) are built. Nucleotides are
organic (p.11) molecules made from aromatic
(p.11) compounds, a simple monosaccharide
(p.43) called ribose (p.46) and phosphate (p.20).

DNA deoxyribonucleic acid. DNA is the main
nucleic acid (↑) of the nucleus (↑). It is a large
macromolecule (p.10) of nucleotides (↑) which
contains the information the cell needs to build
proteins (p.61). This information is contained in
the sequence of nucleotides in each DNA
macromolecule. Most DNA is in the form of a
double helix (p.65) but it is found as a single
chain in some cells, e.g. bacteria (p.148) and
viruses (p.147).

RNA ribonucleic acid. RNA is a single chain
macromolecule (p.10) of nucleotides (↑) which
uses the information stored in DNA (↑) to build
proteins (p.61). *mRNA* (messenger RNA) takes
the information from the nucleus (↑) to the
cytoplasm (p.29). Ribosomes (↓) bond (p.17) to
the mRNA as well as *tRNA* (transfer RNA) to form
polysomes (↓). It is on the polysomes that new
proteins are built.

ribosome (*n*) a small body in the cytoplasm (p.29)
made of ribosomal RNA (↑) and proteins (p.61).

polysome (*n*) a group of ribosomes (↑) bonded
(p.17) to one strand of *mRNA* (↑).

lysosome (*n*) an organelle (p.29) in which
hydrolytic (p.13) enzymes (p.68) are stored.

membrane (*n*) the thin sheet which protects and
encloses cells and organelles (p.29). Made of
protein (p.61) and phospholipid (p.54) it allows
water and other substances to pass through into
and out of the cells and organelles. Cells and
some organelles have a membrane made of a
double layer of phospholipids. **membranous**
(*adj*).

**diagram of the DNA
double helix**

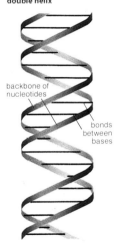

backbone of
nucleotides

bonds
between
bases

cell membrane the double membrane (p.31) surrounding the contents of the cell.

plasmalemma (*n*) = cell membrane (↑).

plasma membrane (*n*) = cell membrane (↑).

nuclear membrane the membrane (p.31) around the nucleus (p.30) of the cell.

endoplasmic reticulum flattened vesicles (p.30) in the cytoplasm (p.29) of cells. They may be rough, with ribosomes (p.31) or smooth, without ribsomes. Newly synthesized (p.13) proteins (p.61) made by the ribosomes enter the endoplasmic reticulum where they may be modified and transported within the cell.

tissue (*n*) any group or collection of cells, e.g. skin, bone, liver (p.41), pancreas (p.29).

pore (*n*) a very small hole through which substances pass. Pores are found in membranes (p.31) and larger surfaces, e.g. skin.

cell wall the stiff wall of a plant cell, on the outside of the cell membrane (↑), usually made of cellulose (p.49). All plants have cell walls but animal cells do not.

respiration (*n*) aerobic (↓) respiration is the process which releases energy from glucose (p.46) and oxygen in the cell. Other molecules, such as lipids (p.51) and amino acids (p.61) can also be used in respiration to produce energy. Glycolysis (↓), the Krebs cycle (p.34) and the electron transfer chain (p.34) are all parts of aerobic respiration and lead to the synthesis (p.13) of ATP (p.36) from ADP (p.36). Anaerobic (↓) respiration does not use oxygen and involves glycolysis and, in some organisms fermentation (p.34) and produces a smaller amount of ATP.
respire (*v*), **respiratory** (*adj*).

aerobic (*adj*) of respiration (↑) which uses oxygen. Also of organisms whose respiration is aerobic.

anaerobic (*adj*) of respiration (↑) which does not use oxygen. Also of organisms whose respiration is anaerobic.

glycolysis (*n*) the chain of anaerobic (↑) reactions (p.12) catalysed (p.68) by enzymes (p.68) where glucose (p.46) is broken down during the first part of respiration (↑) producing pyruvic acid (p.34). *See* diagram opposite.

endoplasmic reticulum

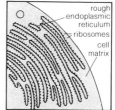

middle lamella

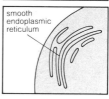

aerobic respiration

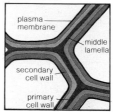

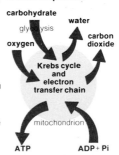

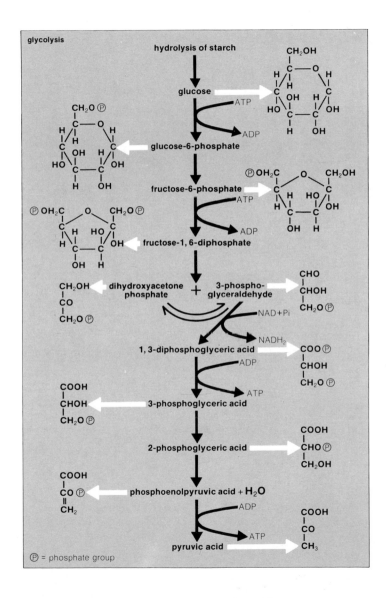

glycolysis

hydrolysis of starch

glucose → ATP → ADP

glucose-6-phosphate

fructose-6-phosphate → ATP → ADP

fructose-1, 6-diphosphate

dihydroxyacetone phosphate + **3-phospho-glyceraldehyde** → NAD+Pi → NADH₂

1, 3-diphosphoglyceric acid → ADP → ATP

3-phosphoglyceric acid

2-phosphoglyceric acid

phosphoenolpyruvic acid + H₂O → ADP → ATP

pyruvic acid

Ⓟ = phosphate group

fermentation (*n*) the breakdown of organic (p.11)
    molecules under anaerobic (p.32) conditions
    especially by bacteria (p.148) or yeast (p.151).
    A well known type of fermentation is in the
    making of alcoholic beverages (p.93) when
    carbon dioxide and the alcohol, ethanol
    (p.21) are produced. **ferment** (*v*).

pyruvic acid $CH_3COCOOH$. A product of glycolysis
    (p.32) and fuel for the Krebs cycle (↓) in aerobic
    (p.32) organisms.

acetyl CoA the final product of glycolysis (p.32).
    Pyruvic acid (↑) is changed to ethanoic acid
    (p.21) and bonded to co-enzyme A (↓) to form
    acetyl CoA. Acetyl CoA feeds into the Krebs
    cycle (↓) to form citric acid (p.21).

co-enzyme A a large molecule formed from ADP
    (p.36) and pantothenic acid (p.77). It acts as the
    co-enzyme (p.71) to the enzyme (p.68) which
    forms citric acid (p.21) from pyruvic acid (↑).

citric acid cycle a cycle which takes place in
    mitochondria (p.30) and is a series of metabolic
    (p.144), enzyme-catalysed reactions (p.12) in
    aerobic (p.32) respiration (p.32) where pyruvic
    acid (↑) is broken down in to carbon dioxide and
    water. The energy released in this process is
    used to produce ATP (p.36) from ADP (p.36) and
    phosphate (p.20) ions (p.9). *See also* citric acid
    (p.21).

TCA cycle tricarboxylic acid cycle = citric acid
    cycle (↑).

Krebs cycle = citric acid cycle (↑). Named after
    Sir Hans Krebs who was the first to describe
    the series of metabolic (p.144) reactions (p.12).
    See diagram opposite.

electron transfer chain the chain of enzyme (p.68)
    redox (p.14) reactions (p.12) which takes place
    in the mitochondrion (p.30). It uses $NADH_2$ (↓)
    from the citric acid cycle (↑) and oxygen to make
    ATP (p.36).

FAD flavine adenine dinucleotide. A carrier of
    hydrogen in the electron transfer chain (↑).

NAD nicotinamide adenine dinucleotide. A carrier
    of hydrogen in the citric acid cycle (↑). $NADH_2$ is
    the reduced (p.14) chemical product which is
    used in the electron transfer chain (↑).

electron transfer system in respiration

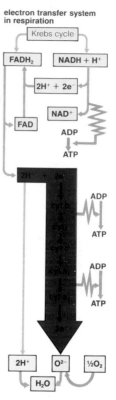

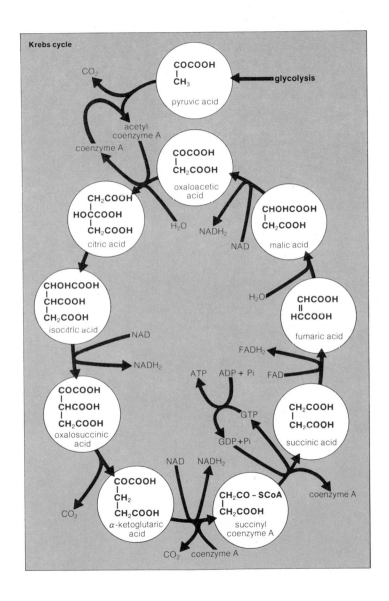

Krebs cycle

ADP adenosine diphosphate. The nucleotide (p.31)
which is formed from the hydrolysis (p.13) of ATP
(↓). It contains two phosphate (p.20) groups.

ATP adenosine triphosphate. The main store of
chemical energy in the cell. It contains three
phosphate (p.20) groups each of which is
bonded (p.17) to the next by a high energy bond.
ATP is the product of respiration (p.32).

hormone (*n*) a substance produced in very small
amounts in the body of a plant or animal which
brings about and/or controls the metabolic
(p.144) activity of cells in other parts of the body.
Hormones are normally made by groups of
special cells known as glands. **hormonal** *(adj)*.

ADP, ATP and their
reactions

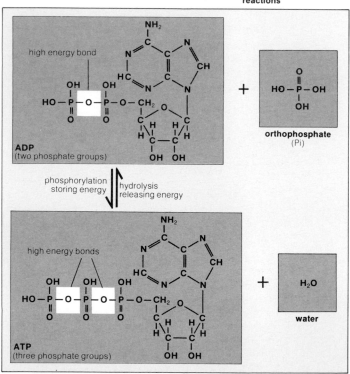

physiology (*n*) the study of the workings of living organisms.

ingestion (*n*) the act of taking food into the body.

mastication (*n*) the breaking down of food in the mouth with the teeth, i.e. chewing. **masticate** (*v*).

saliva (*adj*) the liquid (p.22) which forms in the salivary glands and is released into the mouth. It contains salts (p.16), mucin (↓) and the enzyme (p.68) amylase (↓). Saliva lubricates the particles (p.24) of food and allows them to be swallowed.

mucin (*n*) a viscous (p.23) substance rich in protein (p.61) and carbohydrate (p.43) found in the saliva (↑) which helps in the swallowing of food.

salivary amylase an enzyme (p.68) found in the saliva (↑) which hydrolyses (p.13) starch (p.48) to maltose (p.47).

bolus (*n*) a soft ball of masticated (↑) food formed in the mouth which is swallowed.

oesophagus (*n*) the tube joining mouth to stomach (↓). The walls of the oesophagus contain smooth muscle (p.95) which causes the action known as peristalsis (↓).

peristalsis (*n*) the rhythmical waves of motion in the oesophagus (↑) (and the rest of the alimentary canal (p.42)) caused by contraction (p.97) of smooth muscle (p.95) in the oesophagus wall which moves food from the mouth to the stomach (↓).

stomach (*n*) the large bag-like organ (p.40) into which food passes from the oesophagus (↑). Large amounts of food can be stored in the stomach making it unnecessary for the animal to eat continuously. The food is mixed with gastric juice (p.38) and digestion (p.41) is started. Some food compounds like minerals, glucose (p.46) and vitamins (p.72) may be absorbed (p.41) into the blood through the stomach wall.

section of stomach wall

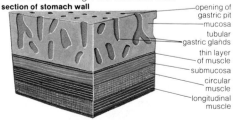

opening of gastric pit
mucosa
tubular gastric glands
thin layer of muscle
submucosa
circular muscle
longitudinal muscle

gastric juice the liquid (p.22) produced by cells in the stomach (p.37) wall. Gastric juice contains the enzyme (p.68) pepsin (↓) and hydrochloric acid (HCl) and has a pH (p.14) of about 2.5–3.0. In infants the gastric juice also contains rennin (↓).

rennin (*n*) enzyme (p.68) found mainly in the gastric juice (↑) in stomachs (p.37) of infants and young animals. Rennin is involved in coagulating (p.67) some proteins (p.61) prior to their digestion by pepsin (↓).

pepsin (*n*) an enzyme (p.68) in the gastric juice (↑) which breaks down proteins (p.61) to peptides (p.63) and amino acids (p.61). It is an acid (p.14) hydrolase (p.69).

chyme (*n*) the partly digested (p.41) food in the stomach (p.37). Chyme consists of a viscous (p.23) liquid (p.22) made of food particles (p.24), gastric juice (↑) and saliva (p.37).

small intestine (*n*) the tube leading out of the stomach (p.37) in which most food digestion (p.41) is done.

**section across
small intestine (ileum)**

muscular wall

villi

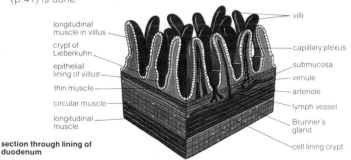

longitudinal muscle in villus

crypt of Lieberkuhn

epithelial lining of villus

thin muscle

circular muscle

longitudinal muscle

section through lining of duodenum

villi

capillary plexus

submucosa

venule

arteriole

lymph vessel

Brunner's gland

cell lining crypt

duodenum (*n*) the first part of the small intestine (↑) leading out of the stomach (p.37). Digestion (p.41) continues in the duodenum through the action of intestinal juice (↓), pancreatic juice (p.40) and bile (p.40).

ileum (*n*) the third and longest part of the small intestine (↑) located after the duodenum (↑). The ileum is muscular (p.95) and causes the movement (by peristalsis (p.37)) of food along its length into the large intestine (p.42).

jejunum (*n*) the part of the small intestine (↑)
between the duodenum (↑) and ileum (↑).

intestinal juice a secretion (p.40) produced by the
small intestine (↑) containing several enzymes
(p.68) which break down food compounds, e.g.
maltase (p.47), lactase (p.47), sucrase and
peptidase (p.69).

pancreas (*n*) an organ (p.40) producing digestive
(p.41) enzymes (p.68) which is connected to the
duodenum (↑). The pancreatic juice (p.40) made
by the pancreas empties into the small intestine
(↑).

diagram of digestive processes and enzymes

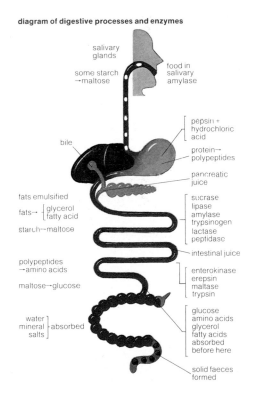

salivary
glands

some starch
→maltose

food in
salivary
amylase

pepsin +
hydrochloric
acid

bile

protein→
polypeptides

pancreatic
juice

fats emulsified

sucrase
lipase
amylase
trypsinogen
lactase
peptidase

fats→ {glycerol
fatty acid

starch→maltose

intestinal juice

polypeptides
→amino acids

enterokinase
erepsin
maltase
trypsin

maltose→glucose

water]
mineral }absorbed
salts]

glucose
amino acids
glycerol
fatty acids
absorbed
before here

solid faeces
formed

organ (*n*) a body of cells in a plant or animal which has a special function.

pancreatic juice the liquid (p.22) secreted (↓) by the pancreas (p.39) containing salts (p.16) which neutralizes (p.16) the acid (p.14) from the stomach (p.37) in the chyme (p.38). It also contains the enzymes trypsin (↓), chymotrypsin (↓), pancreatic amylase (↓) and lipase (↓).

secretion (*n*) a substance, usually a liquid (p.22), with an special function and made inside cells but which passes outside the cell to do its job. Some organs (↑) have cells which produce special secretions, e.g. liver (↓), pancreas (p.39). **secrete** (*v*).

trypsin (*n*) a proteolytic enzyme (p.69) of the pancreatic juice (↑) which breaks down proteins (p.61) into peptides (p.63) and amino acids (p.61). It is an alkaline (p.15) hydrolase (p.69) which attacks different parts of protein molecules to those attacked by pepsin (p.38).

chymotrypsin (*n*) a proteolytic enzyme (p.69) of the pancreatic juice (↑) which breaks down proteins (p.61) into peptides (p.63) and amino acids (p.61). It is an alkaline (p.15) hydrolase (p.69) and attacks different parts of protein molecules to those attacked by pepsin (p.38) and trypsin (↑).

pancreatic amylase a hydrolytic (p.13) enzyme (p.68) of the pancreatic juice (↑) which breaks down starch (p.48) into maltose (p.47). *See* amylose (p.48).

lipase (*n*) a hydrolytic (p.13) enzyme (p.68) of the pancreatic juice (↑) which breaks down triglycerides (p.53) into fatty acids (p.53) and glycerol (p.45).

bile (*n*) the liquid (p.22) which contains the salts (p.16) produced by the liver (↓) which is stored in the gall bladder (↓) and empties into the duodenum (p.38) making the chyme (p.38) slightly alkaline (p.15). It emulsifies (p.84) fats (p.52) making them more easily broken down by lipase (↑).

gall bladder a small, bag-like organ (↑) which stores bile (↑) and is joined to the duodenum (p.38) by the bile duct.

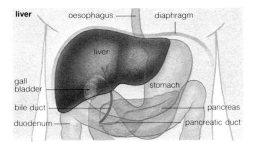

liver, oesophagus, diaphragm, liver, gall bladder, stomach, bile duct, pancreas, duodenum, pancreatic duct

liver (*n*) a large organ (↑) with many important functions. It detoxifies (↓) food compounds secretes (↑) bile (↑), stores iron, vitamins (p.72) A, D and B, excess carbohydrate (p.43) as glycogen (p.50) and synthesizes (p.13) blood proteins (p.61) among other functions.

detoxication (*n*) the breakdown or conversion of toxic (↓) substances in the blood into safe compounds. Detoxication takes place in the liver (↑) **detoxify** (*v*), **detoxification** (*n*).

toxic (*adj*)of substances which are harmful to the body. These toxins can cause illness or death.

poison (*n*) — toxin (↑)

digestion (*n*) the process by which food macromolecules (p.10) are hydrolysed (p.13) in the stomach (p.37) and small intestine (p.38) into small enough molecules that they may be absorbed (↓) and assimilated (↓) into the blood. **digest** (*v*), **digestible** (*adj*).

indigestible (*adj*) of a food or food substance which cannot easily be digested (↑), by humans, e.g. cellulose (p.49), fibre (p.139).

assimilation (*n*) the process by which digested (↑) food is absorbed (↓) through the wall of the small intestine (p.38) into the blood stream. **assimilate** (*v*).

absorb (*adj*) to take up a substance.

absorption (*n*) the active taking up of digested (↑) food nutrients (p.137) through the villi (p.42) of the small intestine (p.38) into blood capillaries (p.42). Absorption is also a more general process in which substances, electromagnetic radiation (p.88) or energy are taken up by other substances.

capillary (*n*) a very small blood vessel.

villus (*n*) a fold in the wall of the small intestine (p.38). Millions of villi (*pl*) increase the surface area of the small intestine making absorption (p.41) much faster.

bowel (*n*) the lower end of the intestine joined to the ileum (p.38). Very little absorption (p.41) of nutrients (p.137) occurs in the bowel but it is here that water is taken up into the blood.

large intestine = bowel (↑).

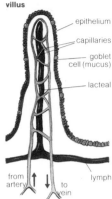

villus

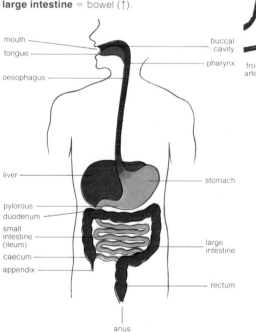

alimentary canal

alimentary canal the oesophagus (p.37), stomach (p.37), duodenum (p.38), jejunum (p.39), ileum (p.38) and large intestine (↑).

gut (*n*) = alimentary canal (↑).

digestive tract = alimentary canal (↑).

carbohydrate (*n*) one of a group of molecules important as nutrients (p.137) which act as a source of energy in the body. Carbohydrates contain carbon, hydrogen and oxygen; some contain other elements (p.8). They exist as single small molecules called monosaccharides (↓) as well as polymers of these called polysaccharides (p.44). Many carbohydrates have aldehyde (−CHO) and ketone (−C=O) functional groups (p.19) and so are reducing (p.14) compounds.

photosynthesis (*n*) the process by which plants make glucose (p.46) and other carbohydrates (↑) from carbon dioxide (CO_2) and water. The reactions (p.12) take place in two stages. (1) Water is hydrolysed (p.13) using the energy of light trapped by chlorophyll (p.30) in the Hill reaction (↓) to produce oxygen and the reducing (p.14) compound $NADPH_2$. (2) The $NADPH_2$ is used in further reactions with ATP (p.36) to produce carbohydrates. These carbohydrates made by plants are the source of all food energy on Earth. **photosynthesize** (*v*).

Hill reaction the reaction (p.12) in green plants in which chlorophyll (p.30) uses the energy of light to hydrolyse (p.13) water. It is the first stage of photosynthesis (↑).

monosaccharide (*n*) a polyhydroxyl (p.44) compound composed mainly of carbon, hydrogen and oxygen. The common monosaccharides such as glucose (p.46) and fructose (p.46) contain six carbon atoms but molecules with three, four, five, seven and more carbon atoms may be found. All polysaccharides (p.44), e.g. starch (p.48), cellulose (p.49), pectin (p.49), are built from monosaccharides.

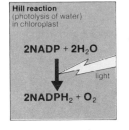

Hill reaction
(photolysis of water)
in chloroplast

$2NADP + 2H_2O$

light

$2NADPH_2 + O_2$

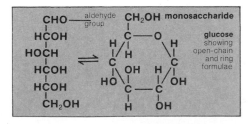

glucose showing open-chain and ring formulae; monosaccharide; aldehyde group

polyhydroxyl (*adj*) of compounds having many hydroxyl (–OH) functional groups (p.19).

disaccharide (*n*) two monosaccharides (p.43) joined together by a glycosidic bond (↓). Sucrose (p.46), lactose (p.47) and maltose (p.47) are examples of important food disaccharides.

disaccharide
e.g. sucrose

CH_2OH

HOCH$_2$

CH_2OH

glycosidic bond

glycosidic bond a covalent bond (p.17) between two carbon atoms in separate monosaccharides (p.43) formed through an oxygen atom.

oligosaccharide (*n*) a short polymer (p.10) chain of monosaccharides (p.43). Oligosaccharides usually contain between two and ten monosaccharides.

polysaccharide (*n*) a polymer (p.10) formed from many monosaccharides (p.43) linked together by glycosidic bonds (↑). Polysaccharides may be in the form of long chains, e.g. glycogen (p.50), or branched chains, e.g. amylopectin (p.48). They act as energy stores in plants and animals as well as being important in the structure of multicellular (p.147) organisms.

homopolysaccharide (*n*) a polysaccharide (↑) built from only one kind of monosaccharide, e.g. starch (p.48).

heteropolysaccharide (*n*) a polysaccharide (↑) built from more than one kind of monosaccharide (p.43).

sugar (*n*) the general name for monosaccharides (p.43) and disaccharides (↑). Common name for the disaccharide sucrose (p.46).

aldose (*n*) the general name for monosaccharides (p.43) with an aldehyde (–CHO) as a functional group (p.19).

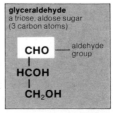

glyceraldehyde
a triose, aldose sugar
(3 carbon atoms)

CHO — aldehyde group
|
HCOH
|
CH_2OH

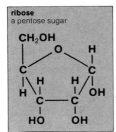

ribose
a pentose sugar

ketose (*n*) the general name for monosaccharides (p.43) with a ketone (−C=O) as a functional group (p.19).

hexose (*n*) a monosaccharide (p.43) with six carbon atoms, e.g. glucose (p.46), fructose (p.46).

pentose (*n*) a monosaccharide (p.43) with five carbon atoms, e.g. ribose (p.46).

pyranose (*n*) a six sided ring formed from a simple monosaccharide (p.43) molecule, e.g. glucose (p.46).

furanose (*n*) a five sided ring formed from a simple monosaccharide (p.43) molecule, e.g. fructose (p.46).

epimer (*n*) an isomer (p.11) of a monosaccharide (p.43) differing only in the way in which the hydroxyl group (−OH) is placed at one of the carbon atoms.

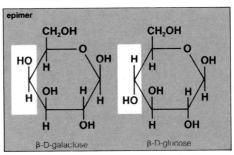

epimer

β-D-galactose   β-D-glucose

epimerase (*n*) an enzyme (p.68) which changes one monosaccharide (p.43) into its epimer (↑).

anomer (*n*) an isomer (p.11) of a monosaccharide (p.43) in which the arrangement of the hydroxyl group (−OH) differs only at the C1 carbon atom.

optical activity the ability of some compounds, particularly sugars (↑), to rotate the plane of polarized light (↓).

polarized light a beam of light waves which vibrate only in one plane.

glycerol (*n*) an aldose (↑) containing only three carbon atoms. Glycerol forms part of the main kind of food lipid (p.51), the triglyceride (p.53).

glycerine (*n*)= glycerol (↑).

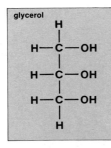

glycerol

glucose (*n*) an aldose (p.44) commonly found in the pyranose (p.45) form. The most common monosaccharide (p.43) and the most important nutrient (p.137) in terms of energy. Glucose is the starting material for glycolysis (p.32). It is also the main product of photosynthesis (p.43).

fructose (*n*) a ketose (p.45) commonly found in the furanose (p.45) form. Fructose is found widely in fruit, especially grapes.

isomerase (*n*) an enzyme (p.68) used in the food industry to convert glucose (↑) to fructose (↑).

galactose (*n*) an aldose (p.44) commonly found in the pyranose (p.45) form. Galactose is often found in the oligosaccharides (p.44) attached to glycoproteins (p.62) and is part of the milk disaccharide (p.44) lactose (↓).

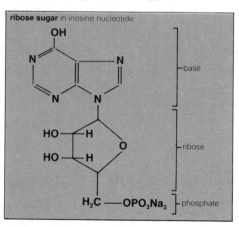

monosaccharide

fructose
showing open-chain and ring formulae

ribose sugar in inosine nucleotide

ribose (*n*) an important pentose (p.45) which is part of the ADP (p.36) and ATP (p.36) molecules. It is also found in nucleotides (p.31) and, therefore, DNA (p.31) and RNA (p.31).

sucrose (*n*) a disaccharide (p.44) formed from glucose (↑) and fructose (↑), commonly known as sugar (p.44). It is extracted (p.90) from sugar cane and sugar beet and has been the most widely used sweetener (p.85) for some centuries.

cane sugar = sucrose (↑).

formation of an invert sugar

invertase (*n*) an enzyme (p.68) used in the food industry to break down sucrose (↑) to glucose (↑) and fructose (↑).

invert sugar a mixture of glucose (↑) and fructose (↑) made by the hydrolysis (p.13) of sucrose (↑).

lactose (*n*) the main sugar (p.44) of milk. A disaccharide (p.44) of glucose (↑) and galactose (↑).

lactase (*n*) a enzyme (p.68) which hydrolyses (p.13) lactose (↑) to glucose (↑) and galactose (↑) found in pancreatic juice (p.40) and also produced by micro-organisms (p.147). Without lactase, humans are unable to digest (p.41) lactose. This condition is known as lactose intolerance (↓) and is widespread and affects about 80% of the world population, especially Africans, Greeks and Orientals.

lactose intolerance a condition, in humans, in which lactose (↑) is not digested (p.41) due to a lack of the enzyme (p.68) lactase (↑).

maltose (*n*) a disaccharide (p.44) of glucose (↑). Maltose is produced when starch (p.48) is hydrolysed (p.13). Also known as **malt sugar.**

maltase (*n*) an enzyme (p.68) found in pancreatic juice (p.40) which breaks maltose (↑) into two molecules of glucose (↑).

reducing sugar a sugar (p.44) or polysaccharide (p.44) with ketone (−C=O) or aldehyde (−CHO) functional groups (p.19) free to react, e.g. glucose (↑), galactose (↑). The functional groups are able to take part in redox (p.14) reactions (p.12) which is why these sugars are called reducing sugars.

hexuronic acid a derivative (p.19) of a hexose (p.45) which contains a carboxylic acid (−COOH) functional group (p.19). An important hexuronic acid is galacturonic acid which is the monomer (p.11) of the pectins (p.49).

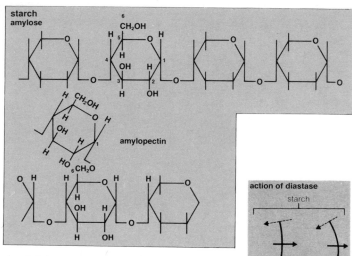

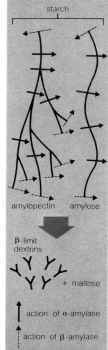

action of diastase

starch

amylopectin   amylose

β-limit
dextrins

+ maltose

↑ action of α-amylase

↑ action of β-amylase

starch (*n*) a polysaccharide (p.44) of glucose
(p.46) found in all plants. Starch is used by
plants as a store of glucose to be to hydrolysed
(p.13) and used in glycolysis (p.32) when
needed. It is found in two forms, amylose (↓) and
amylopectin (↓). When purified and partially
hydrolysed it can be used as a stabilizer (p.84)
or thickener (p.84) in foods.

dextrin (*n*) the short chain form of starch (↑) after
partial hydrolysis (p.13).

amylose (*n*) the straight chain form of the starch
(↑) polymer (p.10). It is quite soluble (p.23) but
forms crystals (p.11) in some conditions, e.g.
starch retrogradation (p.121).

amylopectin (*n*) the branched form of the starch (↑)
polymer (p.10). Amylopectin is insoluble (p.23).

diastase (*n*) a combination of the plant enzymes
(p.68) α- and β- amylase.

gelatinization (*n*) the process by which starch
granules (↓) take up water and swell. Most
starches gelatinize at or near 70°C. Many
starches can form gels (p.25) under the right
conditions on cooling down after gelatinization.
gelatinize (*v*).

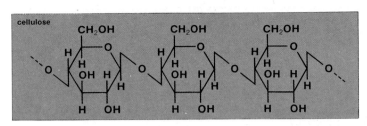

cellulose

common starch granules

potato

rice

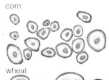

corn

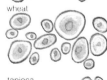

wheat

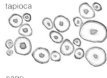

tapioca

sago

starch granule starch (↑) is found in plant cells in the form of grains or granules. These have a different appearance in different plants but generally do not take up cold water.

waxy (*adj*) of the form of starch granule (↑) containing no amylose (↑), e.g. waxy (p.49) maize (p.114).

cellulose (*n*) a carbohydrate (p.43) polymer (p.10) made of glucose (p.46) molecules. Cellulose forms the main structural molecule of the plant cell wall (p.32). It is indigestible (p.41) to humans.

cellulase (*n*) an enzyme (p.68) which breaks down cellulose (↑).

pectin (*n*) a polysaccharide (p.44) of the hexuronic acid (p.47) galacturonic acid found in plant cells and particularly between cells. It interacts (p.16) with cellulose (↑) and other complex polysaccharides to provide a matrix that binds the cells together. Pectin is esterified (p.19) and contains many methoxyl (CH_3O-) functional groups (p.19).

polygalacturonic acid a polysaccharide (p.44) of galacturonic acid. It is the polymer (p.10) known as pectin (↑) without any methoxyl esters (p.19).

pectic acid = polygalacturonic acid (↑).

galacturonic acid

protopectin (*n*) a very large polysaccharide (p.44) from which pectin (p.49) is formed. The word is not now used. Protopectins are macromolecular (p.10) pectins.

pectinase (*n*) an enzyme (p.68) found especially in citrus fruits (p.108) which hydrolyses (p.13) pectins (p.49) into small polygalacturonic acid (p.49) molecules.

pectinesterase (*n*) an enzyme (p.68) which removes the methoxyl ester (p.19) groups (CH_3O-) from pectin (p.49) to give pectic acid (p.49).

hemicellulose (*n*) the name of a group of polysaccharides (p.44) found in plant cell walls (p.32).

glycogen (*n*) a polysaccharide (p.44) of glucose (p.46) similar in structure to amylose (p.48). Glycogen is the main store of glucose in animals and is found in the liver (p.41) and muscles (p.95).

glycoside (*n*) a compound containing a sugar (p.44) bonded (p.17) to another molecule. Glycosides are commonly found in plants and some are toxic.

syrup (*n*) an aqueous (p.23) solution (p.22) of sugars (p.44) which may contain a mixture of monosaccharides (p.43), disaccharides (p.44) and oligosaccharides (p.44).

hygroscopic (*adj*) of any substance which absorbs (p.41) water vapour (p.24) from the air.

caramel (*n*) a brown substance which is made when carbohydrates (p.43) especially sugars (p.44) are heated in solution (p.22). Caramel is used to colour and flavour many foods, e.g. beer (p.93), sauces (p.90). *See also* caramelization (p.92).

honey (*n*) a naturally occuring invert sugar (p.47) produced by bees from plant nectar, which is mainly sucrose (p.46).

uronic acids an important group of sugar (p.44) acids (p.14) in which only the primary hydroxyl ($-OH$) group is oxidized (p.14) to a carboxylic acid ($-COOH$). Common uronic acids found in foods include glucuronic acid and galacturonic acid (p.49).

diagram of the main lipid crystal structures

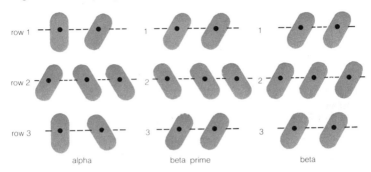

row 1

row 2

row 3

alpha   beta prime   beta

lipid (*n*) the name given to the group of substances
including fats (p.52), oils (p.52), waxes (p.54),
fatty acids (p.53), phospholipids (p.54),
glycolipids (p.55), and steroids (p.54). Lipids are
soluble (p.23) in organic (p.11) liquids (p.22) but
are generally insoluble (p.23) in water.

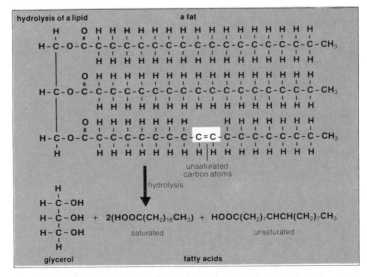

hydrolysis of a lipid   a fat

unsaturated carbon atoms

hydrolysis

$$H-C-OH$$
$$H-C-OH + 2(HOOC(CH_2)_{16}CH_3) + HOOC(CH_2)_7CHCH(CH_2)_7CH_3$$
$$H-C-OH$$
$$H$$

glycerol   saturated   unsaturated   fatty acids

Bloor classification a method of classifying lipids
(p.51) used by Bloor in 1926. He listed four
classes: (1) simple lipids (↓); (2) compound lipids
(↓); (3) glycolipids (p.55); and (4) the derived
lipids (↓).

simple lipids a group of lipids (p.51) including the
triglycerides (↓), esters (p.19) of cholesterol
(p.54) and waxes (p.54).

compound lipids a group of lipids (p.51) most of
which are phospholipids (p.54).

derived lipids a group of lipids (p.51) including
fatty acids (↓), steroids (p.54) and the fat soluble
(p.72) vitamins (p.72) A, D, E and K *(see* pp 72,
73, 74).

lipoid (*adj*) like or similar to a lipid (p.51). Lipoids
are substances which are not simple lipids (↑) or
compound lipids (↑) but are lipid-like, e.g.
steroids (p.54).

fats (*n.pl.*) esters (p.19) of fatty acids (↓) and
glycerol (p.45) which are solid (p.22) at normal
temperatures (18–20°C).

oils (*n.pl.*) esters (p.19) of fatty acids (↓) and
glycerol (p.45) which are liquid (p.22) at normal
temperatures (18–20°C).

fatty acid

saturated

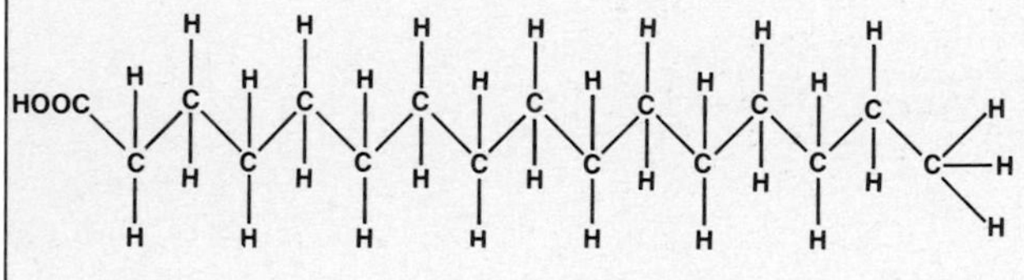

unsaturated

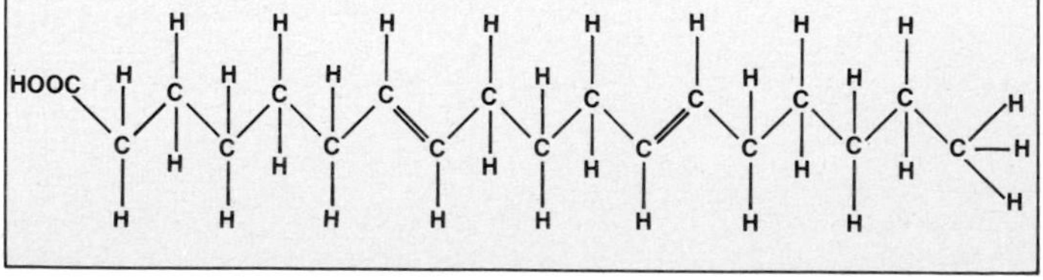

fatty acid a carboxylic acid (–COOH) with a
hydrocarbon (p.11) chain. This chain can contain
few or many carbon atoms but in foods the
number is usually between 4 and 18. Fatty acid
hydrocarbon chains containing no double bonds
(p.18) are said to be saturated (p.18) fatty acids.
Those with double bonds in the hydrocarbon
chain are called unsaturated (p.18) fatty acids.

free fatty acid the name given to fatty acids (↑)
which exist as free compounds not esterified
(p.19) to glycerol (p.45). They are found as
breakdown products of triglycerides (↓) during
digestion (p.41) and after heating lipids (p.51) in
cooking (p.87).

triglyceride (*n*) an ester (p.19) of glycerol (p.45)
containing three fatty acids (↑). Triglycerides are
the most common form of fat (↑) and oil (↑) in
foods. The fatty acids in triglycerides can contain
from 4 to 18 or more carbon atoms.

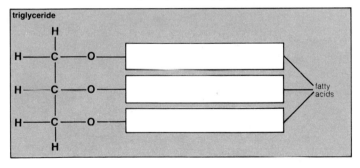

monoglyceride (*n*) an ester (p.19) of glycerol
(p.45) with only one fatty acid (↑).

diglyceride (*n*) an ester (p.19) of glycerol (p.45)
with two fatty acids (↑). Monoglycerides (↑) and
diglycerides are made by the breakdown of
tyriglycerides (↑) during digestion (p.41). They
are also used in the food industry as emulsifiers
(p.84).

polyunsaturated (*adj*) of fatty acids (↑), fats (↑) or
oils (↑) that contain many double bonds (p.18)
between carbon atoms in the hydrocarbon (p.11)
chains.

cholesterol

CH_3

CH — CH_2 — CH_2 — CH_2 — CH

CH_3

CH_3

HO

cholesterol (*n*) the main sterol (↓) in animal tissues
(p.32). It is made up of four joined carbon rings
and a short hydrocarbon (p.11) chain. High levels
of cholesterol in the diet (p.137) have been linked
with heart disease and atherosclerosis (p.140).

steroid (*n*) one of a group of lipids (p.51) which do
not contain glycerol (p.45). Steroids have four
rings of carbon atoms joined together and often
several functional groups (p.19) as well as
hydrocarbon (p.11) chains.

sterol (*n*) a steroid (↑) with a hydroxyl (–OH)
functional group (p.19) at carbon atom 3.

phytosterol (*n*) a steroid (↑) found in vegetable oils
(p.52).

wax (*n*) a group of simple lipids (p.52) most of
which are esters (p.19) of large alcohols and fatty
acids (p.53). Waxes are solid (p.22) and are most
commonly found in foods as the waterproof
coating on the skin of vegetables and fruit.

phospholipid (*n*) a class of compound lipids (p.52)
which are made up of glycerol (p.45) and fatty
acids (p.53) but also functional groups (p.19)
containing phosphorus. Phospholipids form the
structure of most cell and organelle (p.29)
membranes (p.31) and so are found in all natural
foods.

phospholipid lecithin

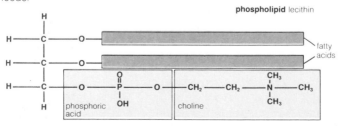

fatty
acids

phosphoric
acid

choline

lecithins (*n*) a group of phospholipids (↑) which contain phosphoric acid (–HPO$_4^-$)and choline (–CH$_2$CH$_2$N$^+$[CH$_3$]$_3$). The lecithins are found in many foods including egg yolk, yeasts (p.151), legumes (p.109), cereals (p.112) and animal tissues (p.32) such as liver (p.41). They are important as natural emulsifying agents (p.84).

glycolipids (*n*) a group of lipids (p.51) which contain carbohydrates (p.43). Glycolipids are found in cell and organelle (p.29) membranes (p.31) and nerve tissue (p.32).

tallow (*n*) extracted (p.90) or rendered (↓) animal fat (p.52).

lard (*n*) extracted (p.90) or rendered (↓) pig fat (p.52).

render (*adj*) to remove the fat (p.52) from an animal carcass (p.98) by heating. This process melts (↓) the fat and allows it to be collected.

melting point the temperature at which a solid (p.22) turns into a liquid (p.22). As most triglycerides (p.53) are heterogeneous (p.24) and contain a mixture of fatty acids (p.53) they melt over a range of temperatures. Most fats (p.52) melt between 30°C and 40°C. **melt** (*v*).

slip point the temperature at which a triglyceride (p.53) begins to melt (↑). The slip point can be measured by melting the fat (p.52) or oil (p.52) in a glass tube and noting the temperature at which the solid (p.22) begins to slide.

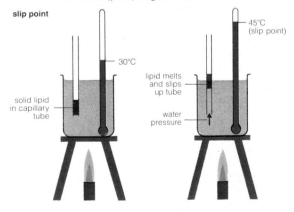

slip point

30°C

solid lipid in capillary tube

45°C (slip point)

lipid melts and slips up tube

water pressure

smoke point the temperature at which a fat (p.52)
or oil (p.52) begins to smoke. This smoke may be
seen as a blue haze over the hot fat or oil and
marks the beginning of breakdown of the lipid
(p.51).

acrolein (*n*) propenal or acraldehyde
(CH_2=CHCHO). Acrolein is formed when glycerol
(p.45) is heated to high temperatures. At the
smoke point (↑), triglycerides (p.53) are broken
down to fatty acids (p.53) and glycerol which
itself breaks down to acrolein.

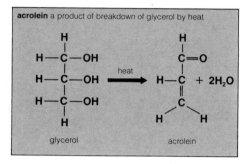

acrolein a product of breakdown of glycerol by heat

glycerol   heat →   acrolein   + $2H_2O$

flash point the temperature at which fat (p.52) or
oil (p.52) will immediately ignite (↓).

ignite (*v*) to burst into flames.

iodine number the number of grams of iodine
absorbed (p.41) by 100 grams of fat (p.52) or oil
(p.52). The iodine number gives a measure of the
degree of unsaturation (p.18) of the fat or oil as
iodine reacts (p.12) with double bonds (p.18) in
the fatty acids (p.53).

fat crystals crystals (p.11) of fats (p.52) and oils
(p.52) formed under some conditions. The kinds
of crystal formed in a lipid (p.51) can affect its
properties, e.g. melting point (p.55), aeration (↓).

polymorphism (*n*) the ability of some triglycerides
(p.53) to exist as several different kinds of
crystal (p.11). The three forms are known as α, β
and γ.

bleaching (*n*) a stage in the production of lipids
(p.51) in which colloidal (p.25) material and
coloured compounds are removed.

deodorizing (*n*) a stage, after bleaching (↑), in the production of fats (p.52) and oils (p.52) when flavoured compounds are removed.

wintering (*n*) the removal of more saturated (p.18) triglycerides (p.53) from oils (p.52). The oil is cooled and the resulting solid (p.22) crystals (p.11) are filtered off. The oil then stays clear at low temperature and does not form a haze of fine crystals.

soap (*n*) a substance used for washing made from alkali (p.15) and fat (p.52) or oil (p.52). The salt (p.16) of a fatty acid (p.53).

saponification (*n*) the process by which soap (↑) is made. Alkali (p.15) reacts (p.12) with triglycerides (p.53) to give glycerol (p.45) and the salts (p.16) of fatty acids (p.53) which are called soaps.

creaming (*n*) the mixing of a fat (p.52) or oil (p.52) until it is smooth and creamy.

aeration (*n*) the mixing of small air bubbles with a substance, for example, with a fat (p.52) or oil (p.52) during the process of creaming (↑). **aerate** (*v*).

plasticity (*n*) the ability of a solid (p.22) to change shape when pressure is applied and to remain in the new shape when the pressure is removed. Fats (p.52) are plastic at certain temperatures and can be spread or mixed with other foods easily.

interesterification (*n*) the process by which fatty acids (p.53) are removed from one triglyceride (p.53) and added to an alcohol. Monoglycerides (p.53) and diglycerides (p.53) are often made in this way in the food industry.

superglycerinated fat a mixture of diglycerides (p.53) and monoglycerides (p.53) made by interesterification (↑) of one fat (p.52) to glycerol (p.45).

lypolysis (*n*) the breakdown of triglycerides (p.53) by removal of the fatty acids (p.53) from the glycerol (p.45). Lypolysis is usually carried out by an enzyme (p.68), e.g. a lipase (p.40).

reversion (*n*) a process in which some oils (p.52) which contain the fatty acid (p.53) linolenic acid (e.g. soybean oil) are oxidized (p.14).

auto-oxidation (*n*) the process by which
  triglycerides (p.53) are oxidized (p.14). It is
  called auto-oxidation because, once started, the
  process catalyses (p.68) itself. It has three
  stages, (1) initiation (↓), (2) propagation (↓) and
  (3) termination (↓).

auto-oxidation lipid oxidation

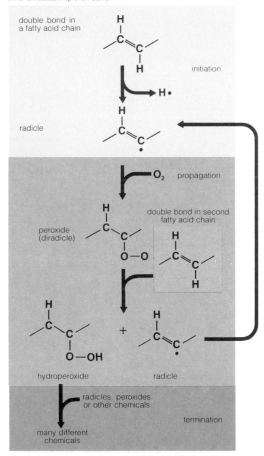

initiation (*n*) the first stage of lipid (p.51) auto-oxidation (↑). Initiation is the oxidation (p.14), by the removal of hydrogen, of a fatty acid (p.53) chain in a triglyceride (p.53) to form a radicle (↓).

radicle (*n*) a highly reactive (p.12) derivative (p.19).

propagation (*n*) the second stage in lipid (p.51) auto-oxidation (↑). Oxygen is added to the radicle (↑) formed in the initiation (↑) stage to form a peroxide. This compound can form a hydroperoxide by removing a hydrogen atom from nearby fatty acid (p.53). This process therefore forms a new radicle and in this way the process is self-continuing.

termination (*n*) the third stage in lipid (p.51) auto-oxidation (↑). The hydroperoxide formed in propagation (↑) reacts (p.12) with oxygen and other compounds. The products of these reactions often smell and taste bad.

rancidity (*n*) the bad smell and taste in fats (p.52) and oils (p.52) after auto-oxidation (↑) or lypolysis (p.57). **rancid** (*adj*).

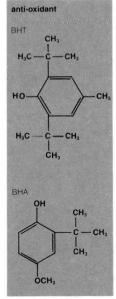

anti-oxidant

BHT

BHA

lipoxidase (*n*) an enzyme (p.68) which oxidizes (p.14) many triglycerides (p.53) to hydroperoxides. It is used in the baking industry to bleach (p.121) flour (p.116).

lipoxygenase (*n*) = lipoxidase (↑).

anti-oxidant (*n*) a compound added to a lipid (p.51) or any other food to stop oxidation (p.14). Anti-oxidants are added to a large number of foods to help lengthen their storage lives by inhibiting (p.70) rancidity (↑).

BHT butylated hydroxytoluene. A common food anti-oxidant (↑).

BHA butylated hydroxyanisole which is used as an anti-oxidant (↑) in many foods.

butylated (*adj*) of a chemical which has had a butyl ($CH_3CH_2CH_2CH_2-$) functional group (p.19) added to form a butyl derivative (p.19). **butylation** (*n*), **butylate** (*adj*).

propyl gallate a common anti-oxidant (↑).

chelating reagents compounds which form bonds (p.17) with metal ions (p.9). Metal ions act as catalysts (p.68) for lipid (p.51) oxidation (p.14) but this can be stopped by bonding such ions to chelating reagents. *See* chelator (p.86).

synergist (*n*) a compound which increases the effectiveness of an anti-oxidant (p.59), e.g. some chelating reagents (p.59) are also synergists (citrate, phosphate).

hydrogenation (*n*) the treatment of unsaturated (p.18) lipids (p.51) with hydrogen to make them more saturated (p.18). The process is done with catalysts (p.68) and under pressure. Hydrogenation is used with oils (p.52) to make them solid (p.22) enough for use in the production of margarine (↓).

margarine (*n*) a creamy substance usually made from hydrogenated (↑) vegetable oils (p.52), emulsifiers (p.84), colours and added vitamins (p.72). Margarine is a food analogue (p.91) of butter (↓) but has a much lower saturated (p.18) fatty acid (p.53) content (about 20% by weight). Most margarines are about 70–80% fat.

votator (*n*) a machine used to cool and solidify margarine (↑) emulsion (p.26) and work it into a semi-solid (p.22) condition ready for packing.

butter (*n*) solid fat (p.52) made from the lipid (p.51) part of milk. Butter is about 80% fat and contains a high proportion of saturated (p.18) fatty acids (p.53) (about 46% by weight).

shortening (*n*) a fat (p.52) or oil (p.52) used to produce a particular effect in baked foods, e.g. butter (↑), margarine (↑) and lard (p.55). Different shortenings can be used to make crisp, soft or flaky pastry, cakes and bread.

enrobing

centre

enrobing
chocolate

enrobing (*n*) the covering of a food with a coating. Usually used of chocolate covered goods.

bloom (*n*) the white surface on chocolate which is seen after incorrect or overlong storage. Bloom is caused by a change in form of the fat (p.52) in the chocolate. Bloom is also the yeast (p.151) which naturally grows on the surface of grapes.

protein (*n*) organic (p.11) macromolecules (p.10) made of amino acids (↓) found in all living cells. There are many different types of protein, each with its own sequence of amino acids. Some are structural proteins, e.g. in membranes (p.31) and some are enzymes (p.68). Proteins are needed in the diet (p.137) as well as lipids (p.51), carbohydrates (p.43), vitamins (p.72), water and minerals (p.16). Many foods such as meat (p.95), fish, beans, pulses (p.109) and eggs have a high protein content. Proteins are the main source of nitrogen in the diet and form the main part of the nitrogen cycle (p.20).

simple proteins proteins (↑) formed only from amino acids (↓).

conjugated proteins proteins (↑) made of amino acids (↓) in chains and also other compounds, e.g. nucleic acids (p.31) in nucleoproteins (p.62), lipids (p.51) in lipoproteins (p.62), carbohydrates (p.43) in glycoproteins (p.62). The non-protein part is called the prosthetic group (p.62).

amino acids a class of organic (p.11) compounds with a carboxyl group (–COOH), an amino group (–NH$_2$), a hydrogen atom and a side group all attached to a central carbon atom. Different amino acids have different side chains (↓). About twenty different amino acids are commonly found in proteins (↑) which they form when joined together by peptide bonds (p.64).

side chain an atom or group of atoms with different functional groups (p.19) bonded (p.17) to the central carbon atom of the amino acid (↑).

amino acids

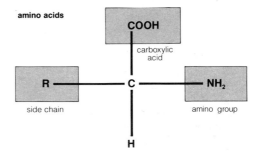

COOH

carboxylic
acid

R — C — NH$_2$

side chain   amino group

H

zwitterion (*n*) a form of an amino acid (p.61) in which its amino group (–NH$_2$) carries a positive charge (–NH$_3^+$) and its carboxylic acid group (–COOH) carries a negative charge (–COO$^-$).

terminal amino acid the end amino acid (p.61) of each polypeptide (↓) chain where there is always one *N-terminal amino acid* with a free amino group (–NH$_2$) and one *C-terminal amino acid* with a free carboxyl group (–COOH).

chromoprotein (*n*) a conjugated protein (p.61) containing a prosthetic group (↓) that gives colour, e.g. myoglobin (p.79) and chlorophyll (p.30).

prosthetic group a non-amino acid compound attached to a protein (p.61), often containing a metal atom such as iron (Fe) or magnesium (Mg).

lipoprotein (*n*) a compound of protein (p.61) and lipid (p.51). Lipoproteins are found in the blood and help to carry lipids throughout the body.

metalloprotein (*n*) a protein (p.61) containing one or more atoms of a metal element (p.8), e.g. iron (Fe), copper (Cu), magnesium (Mg). Ferritin, which contains Fe(OH)$_3$ and the enzyme (p.68) alcohol dehydrogenase are examples of metalloproteins.

nucleoprotein (*n*) a protein (p.61) containing or bonded (p.17) to nucleic acids (p.31) and usually found in the nucleus (p.30) of the cell.

phosphoprotein (*n*) a protein (p.61) containing one or more phosphate (p.20) groups, e.g. casein (p.125) in milk and the enzyme (p.68) pepsin (p.38)

glycoprotein (*n*) a protein (p.61) bonded (p.17) to a sugar (p.44) or to an oligosaccharide (p.44). Most proteins secreted (p.40) from cells are glycoproteins, e.g. mucin (p.37), γ-globulin.

phosphoglycoprotein (*n*) a glycoprotein (↑) which also contains phosphate (p.20).

contractile protein a protein (p.61) which can contract (p.97), e.g. actomyosin (p.96) in muscle (p.95).

lectin (*n*) a protein (p.61) common in plants and found in some animal tissues (p.32). Some lectins have been found in beans and pulses (p.109) like red kidney beans. Lectins can cause illness if not fully denatured (p.66) by cooking (p.87).

isoelectric point the pH (p.14) at which a protein
(p.61) is neutral, i.e. carries no electric charge.
essential amino acids there are 8 essential amino
acids (p.61) which must be supplied by proteins
(p.61) in the diet (p.137) as they cannot be made
from any other food in the body. These essential
amino acids are lysine, methionine, valine,
tryptophan, leucine, isoleucine, threonine, and
phenylalanine. Arginine and histidine may be
needed for growth in children. Some proteins
have a balanced content of the essential amino
acids and are therefore said to have a high
biological value (p.67). Examples of such
proteins are found in eggs, dairy products, fish
and meat. Proteins with a low biological value
contain a less balanced content of essential
amino acids, e.g. collagen (p.98), which contains
only small quantities of histidine and no
tryptophan, and some plant proteins.

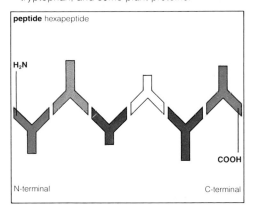

peptide hexapeptide

H_2N

COOH

N-terminal   C-terminal

peptide (*n*) a compound made from two (a
dipeptide) or more amino acids (p.61) joined in a
chain.
oligopeptide (*n*) a peptide (↑) containing between
6 and 50 amino acids (p.61).
polypeptide (*n*) a peptide (↑) with a large number
of amino acids (p.61) in a chain. Polypeptide
chains become folded to form proteins (p.61).

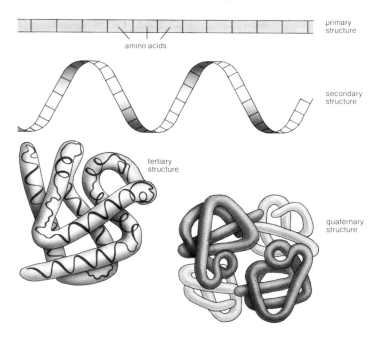

formation of a peptide bond

peptide bond a bond (p.17) formed between the
  amino group (–NH$_2$) of one amino acid (p.61),
  and the carboxyl group (–COOH) of another.

primary, secondary, tertiary and quaternary structure of proteins

primary
structure

amino acids

secondary
structure

tertiary
structure

quaternary
structure

disulphide bond
globular protein

disulphide bond

fibrous protein

disulphide bond

protein structure proteins (p.61) have four levels of structure – primary, secondary, tertiary, quaternary. The *primary structure* is the sequence of the amino acids (p.61) in a polypeptide (p.63). The *secondary structure* of a protein is the coiling of the polypeptide. The *tertiary structure* of a protein is the twisting and folding of the coiled polypeptide to form a three-dimensional protein molecule. The *quaternary structure* of a protein is the structure of several protein molecules when bonded (p.17) together.

helix (*n*) a coil. A common structure in proteins (p 61) **helical** (*adj*).

native conformation the shape or form in which proteins (p.61) are found in living cells. Proteins must be in their native conformation to carry out their functions.

disulphide bond a chemical bond (p.17) formed between the same or different polypeptide (p.63) chains. The bond is made between two sulphydryl groups (–SH) of the amino acid (p.61), cysteine (–S–S–).

globular (*adj*) of a protein (p.61) having a round, ball-like shape. Most globular proteins are soluble (p.23), e.g. enzymes (p.68).

fibrous (*adj*) of a protein (p.61) having a long, string-like shape. Fibrous proteins are often insoluble (p.23) and are used in the body for strength in connective tissues (p.32), e.g. tendons, ligaments, skin, bones and teeth.

globular proteln
myoglobin

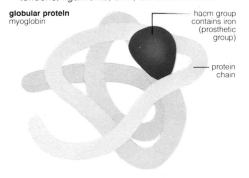

hacm group contains iron (prosthetic group)

protein chain

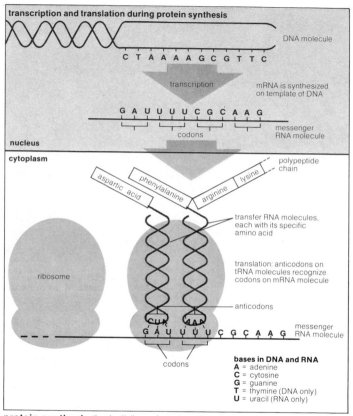

transcription and translation during protein synthesis

DNA molecule

C T A A A A G C G T T C

transcription

mRNA is synthesized on template of DNA

G A U U U U C G C A A G

codons

messenger RNA molecule

nucleus

cytoplasm

polypeptide chain

aspartic acid

phenylalanine

arginine

lysine

transfer RNA molecules, each with its specific amino acid

translation: anticodons on tRNA molecules recognize codons on mRNA molecule

ribosome

anticodons

CUA AAR

messenger RNA molecule

G A U U U U C G C A A G

codons

bases in DNA and RNA
A = adenine
C = cytosine
G = guanine
T = thymine (DNA only)
U = uracil (RNA only)

protein synthesis the building of chains of amino acids (p.61) which takes place in the ribosomes (p.31) of a cell.

denature (v) to bring about changes in protein (p.61) structure, often with loss of function, including the loss of native conformation (p.65). Heat, violent shaking, changes in pH (p.14) and ultraviolet light can make proteins denature. Sometimes these changes can be reversed to give the native conformation. **denaturation** (n).

denaturation
native conformation of protein

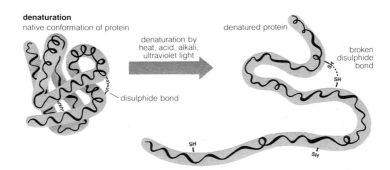

denaturation by
heat, acid, alkali,
ultraviolet light

denatured protein

broken
disulphide
bond

disulphide bond

coagulate (*v*) to change the structure of a protein
(p.61) such that the denaturation (↑) cannot be
changed back and the protein is made insoluble
(p.23), e.g. when egg white is cooked.

biological value a measure of the nutritive (p.137)
value of a protein (p.61) food. It is the amount of
absorbed (p.41) protein which is made into body
protein by protein synthesis (↑).

net protein utilization the amount of protein (p.61)
eaten, which is made into body protein, e.g.
muscle (p.95), enzymes (p.68) and fibrous (p.65)
proteins.

complementary proteins proteins (p.61) which
separately lack certain essential amino acids
(p.68) but when eaten together provide a good
balance of these amino acids (p.61). Many plant
proteins are lacking in some of the essential
amino acids and so, on their own, cannot provide
enough of these amino acids (p.13) of new protein.

textured vegetable protein a protein (p.61) food
made from vegetables which is treated to appear
like meat (p.95).

single cell protein proteins (p.61) made from
single cell organisms such as bacteria (p.148),
moulds (p.151), algae (p.147) and yeasts (p.151)
which are used to make food.

leaf protein proteins (p.61) made from the leaves
of food crops like beans, potatoes, and sugar
beet, which are normally thrown away. These
may be used in foods.

the function of enzymes in catalysis of reactions

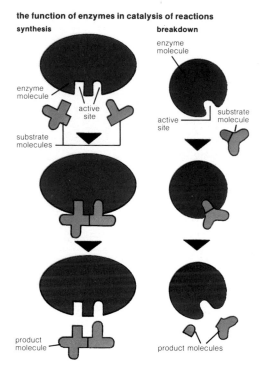

synthesis

breakdown

enzyme molecule

enzyme molecule

active site

substrate molecule

active site

substrate molecules

product molecule

product molecules

enzyme (*n*) a protein (p.61) catalyst (p.68). Small amounts of enzymes are able to speed up chemical reactions (p.62) in cells often by many thousands of times. Each enzyme acts on a different compound and each cell contains many thousands of enzymes. There are several groups of enzymes such as the synthetases (↓), the oxidoreductases (↓), the isomerases (↓) and the hydrolases (↓).

catalyst (*n*) a substance which speeds up the rate of a chemical reaction (p.12) without being changed by the reaction itself.

catalysis (*n*) the process by which a catalyst (↑) speeds up a chemical reaction (p.12). **catalyse** (*v*), **catalytic** (*adj*).

synthetases (*n.pl.*) a group of enzymes (↑) which catalyse (↑) redox (p.14) reactions (p.12).

ligases (*n.pl.*) = synthetases (↑).

oxidoreductases (*n.pl.*) a group of enzymes (↑) which catalyse (↑) redox (p.14) reactions (p.12).

isomerases (*n.pl.*) a group of enzymes (↑) which catalyse (↑) reactions (p.12) in which only the arrangement of atoms in the substrate (p.70) molecule is changed.

hydrolases (*n.pl.*) a group of enzymes (↑) which catalyse (↑) hydrolytic (p.13) reactions (p.12).

protease (*n*) a hydrolase (↑) which breaks peptide bonds (p.64) and forms peptides (p.63) and amino acids (p.61).

proteolytic enzyme = protease (↑).

peptidase (*n*) = protease (↑).

exopeptidase (*n*) a protease (↑) which breaks peptide bonds (p.64) only at the end of the protein (p.61) chain.

exoprotease (*n*) = exopeptidase (↑).

endopeptidase (*n*) a protease (↑) which breaks peptide bonds (p.64) at any point in the protein (p.61) chain.

endoprotease (*n*) = endopeptidase (↑).

proteolysis
by proteolytic enzymes

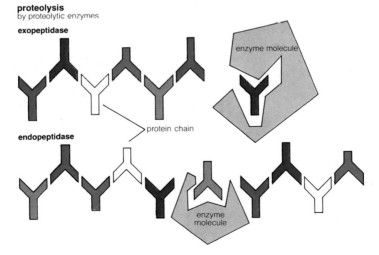

exopeptidase

enzyme molecule

protein chain

endopeptidase

enzyme molecule

enzyme activity a measurement of how an enzyme (p.68) works and how fast it works.

active site the part of the enzyme (p.68) molecule which interacts (p.16) with the substrate (↓) and where the catalysis (p.68) of the reaction (p.12) takes place. The active site usually has a special shape so that the substrates can bond (p.64) to it.

substrate (*n*) a compound which interacts (p.16) with the active site (↑) of an enzyme (p.68) and which is changed during the reaction (p.12).

product[1] (*n*) a compound produced during the reaction (p.12) catalysed (p.68) by an enzyme (p.68). *See also* product (p.168).

inhibitor (*n*) a compound which stops the reaction (p.12) catalysed (p.68) by an enzyme (p.68). Inhibitors often act by interacting (p.16) with active site (↑) in place of the substrate (↑). **inhibit** (*v*).

inhibition (*n*) the process in which the action of an enzyme (p.68) is slowed down or stopped. Reactions (p.12) which are not catalysed (p.68) by enzymes can also be inhibited.

irreversible inhibition inhibition (↑) of the activity (↑) of an enzyme (p.68) which is final and cannot be reversed.

reversible inhibition inhibition (↑) of the activity (↑) of an enzyme (p.68) which can be reversed. There are two types of reversible inhibition, competitive inhibition (↓) and non-competitive inhibition (↓).

competitive inhibition inhibition (↑) of the activity (↑) of an enzyme (p.68) by a chemical which is like the normal substrate (↑) of that enzyme and which can compete with the substrate in binding to the enzyme. Competitive inhibition can be reversed by increasing the concentration (p.22) of the substrate.

non-competitive inhibition inhibition (↑) of the activity (↑) of an enzyme (p.68) by a chemical which does not compete with the substrate (↑) in binding to the enzyme but binds to a part of the enzyme other than the active site (↑). Non-competitive inhibition cannot be reversed by changing the concentration (p.22) of the substrate but only by removing the inhibitor (↑).

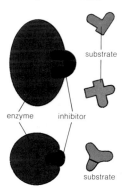

non-competitive inhibition

substrate

enzyme   inhibitor

substrate

co-factor (n) a compound needed by many enzymes (p.68) to carry out the reactions (p.12) which they catalyse (p.68). Many vitamins (p.72) are co-factors to enzymes.

co-enzyme (n) = co-factor (↑).

holoenzyme (n) the molecule formed when a co-factor (↑) interacts (p.16) with an enzyme (p.68).

reversible reaction all enzyme (p.68) catalysed (p.68) reactions (p.12) are reversible, i.e. products can be made from substrates (↑) and substrates can be made from products.

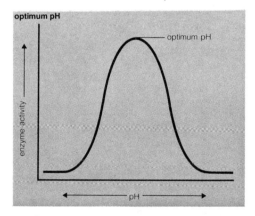

optimum pH

optimum pH

enzyme activity

pH

optimum pH the pH (p.14) at which an enzyme (p.68) is most active (↑).

optimum temperature the temperature at which an enzyme (p.68) is most active (↑). Above optimum, enzymes may be denatured (p.66) and below the rate may be slowed considerably.

isoenzyme (n) a different form of an enzyme (p.68). Isoenzymes of an enzyme have different structures and may be found in different tissues (p.32) but they all catalyse (p.68) the same reaction (p.12).

isozyme (n) = isoenzyme (↑).

zymogen (n) the inactive form of an enzyme (p.68). Many of the hydrolytic (p.13) enzymes of the digestive tract (p.42) are secreted (p.40) as zymogens and need to be partly broken down by other proteases (p.69) before they are able to act.

vitamin (*n*) an organic (p.11) compound needed in very small amounts by living cells. Many vitamins are used in cells as co-enzymes (p.71) in metabolic (p.144) reactions (p.12). Because of this, they may be involved in regulating enzyme (p.68) reaction, reproduction (p.107) and repair of cells and, therefore, generally promoting good health. Vitamins are found in many different foods, especially green vegetables, fruit, meat, eggs, fish and milk. Although they are not related to one another, the vitamins can be divided into two classes, the fat soluble (↓) vitamins and the water soluble (↓) vitamins. *See also* appendix four pp. 172–73.

water soluble of vitamins (↑) which are soluble (p.23) in water, e.g. vitamin C (p.74), niacin (p.76), thiamin (p.75).

leaching (*n*) the loss of water soluble (↑) vitamins (↑) from a food during preparation, cooking (p.87) or blanching (p.89).

fat soluble of vitamins (↑) which are soluble (p.23) in lipids (p.51) or organic (p.11) solvents (p.22), e.g. vitamin A (↓), vitamin D (↓), vitamin E (p.74), vitamin K (p.74).

vitamin deficiency a lack of the correct amount of one or more vitamins (↑) in the diet (p.137). Because vitamins are needed for so many reactions (p.12) in all cells in the body, vitamin deficiency often leads to illness.

vitamin excess an intake of too much of one or more vitamins (↑) in the diet (p.137). In most cases this is not harmful but an excess of vitamins A (↓) and D (↓) can lead to illness.

retinol (*n*) vitamin A. Retinol is a fat soluble (↑) vitamin (↑) which is pale yellow in colour and is found in fish liver (p.41) oils (p.52), animal liver, butter (p.60), margarine (p.60) cheese and eggs. It can be formed from other, similar compounds such as carotene (p.80). Retinol deficiency (↑) leads to night blindness (↓), a slow rate of growth in children, skin disease and a disease of the eye called xerophthalmia (↓).

night blindness poor sight in dark conditions.

xerophthalmia (*n*) a disease of the eye in which the surface of the eye becomes dry and sore.

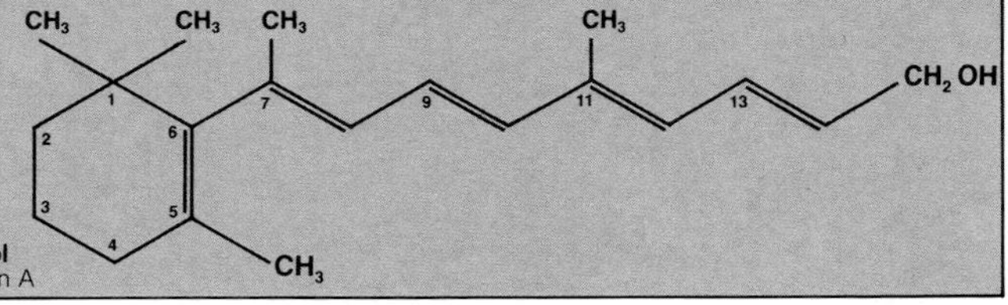

retinol
vitamin A

cholecalciferol (*n*) vitamin D_1. Cholecalciferol is a
fat soluble (↑) vitamin (↑) found in a few foods
like fish liver (p.41) oil (p.52), margarine (p.60),
butter and eggs. It is formed in the skin of
humans on exposure to sunlight. People with
dark skins make less of the vitamin than those
with pale skins. Deficiency (↑) of cholecalciferol
in children leads to the disease called rickets (↓)
and in adults the disease called osteomalacia (↓).

rickets (*n*) a disease in children resulting from a
lack of vitamin D (↑) in the diet (p.137). The
bones do not take up enough calcium and are
soft and bent. This can cause bow legs in some
young children or a deformed pelvis which can
cause problems during childbirth.

osteomalacia (*n*) a disease resulting from a lack of
vitamin D in the diet (p.137) which is common in
old people. Osteomalacia leads to a softening of
the bones and they become weak and painful
and may fracture.

ergocalciferol (*n*) a synthetic (p.13) organic (p.11)
compound known as vitamin D_2. Ergocalciferol is
added to many foods such as margarine (p.60)
and baby foods.

ergocalciferol vitamin D_2

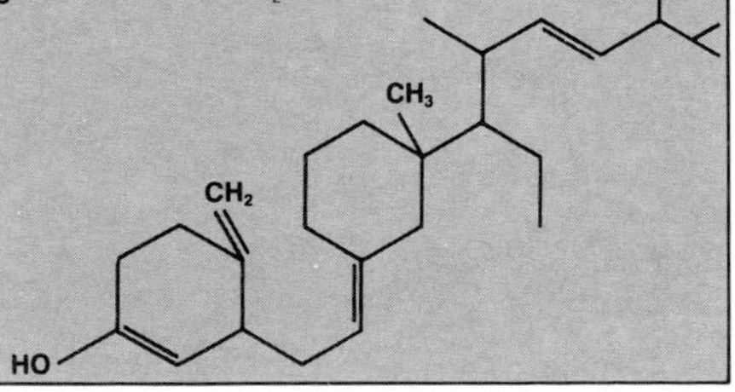

α-**tocopherol** vitamin E

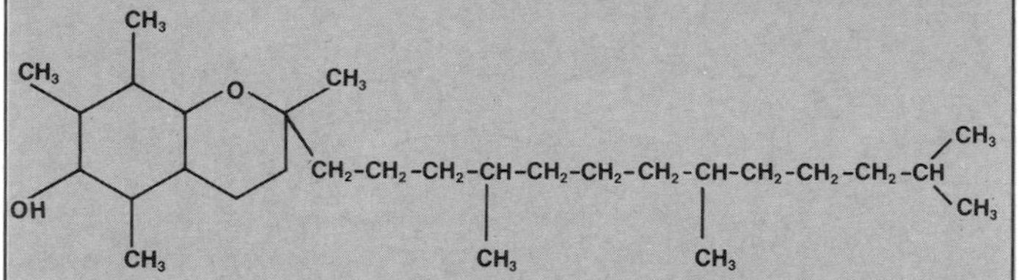

vitamin C
L-ascorbic acid

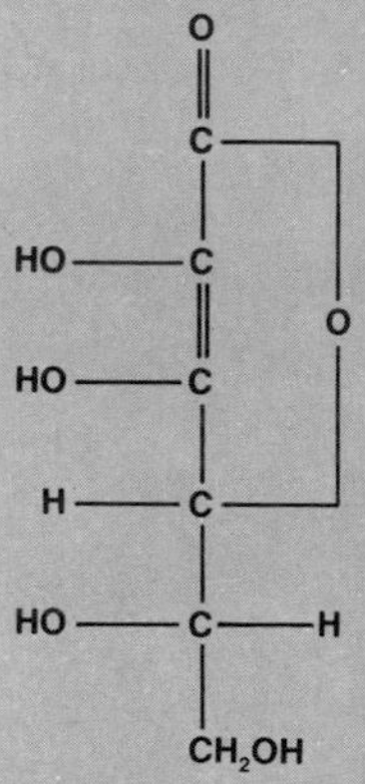

tocopherol (*n*) vitamin E. Tocopherol is a fat soluble (p.72) vitamin (p.72) found in many foods including vegetable oils (p.52), wheat germ (p.108), eggs and milk. Vitamin E is not linked to any known deficiency (p.72) diseases but is a powerful natural anti-oxidant (p.59).

vitamin K phylloquinone. It is a fat soluble (p.72) vitamin (p.72) found in green vegetables. Deficiency (p.72) leads to increased blood clotting times.

vitamin C ascorbic acid. It is a water soluble (p.72) vitamin (p.72) found in blackcurrants, citrus fruits (p.108), potatoes, turnips and many green vegetables. A deficiency (p.72) of vitamin C leads to scurvy (↓).

scurvy (*n*) a disease caused by lack of vitamin C (↑) in the diet (p.137) which leads to anaemia (p.77), lack of energy and deterioration (p.155) of the gums.

vitamin K

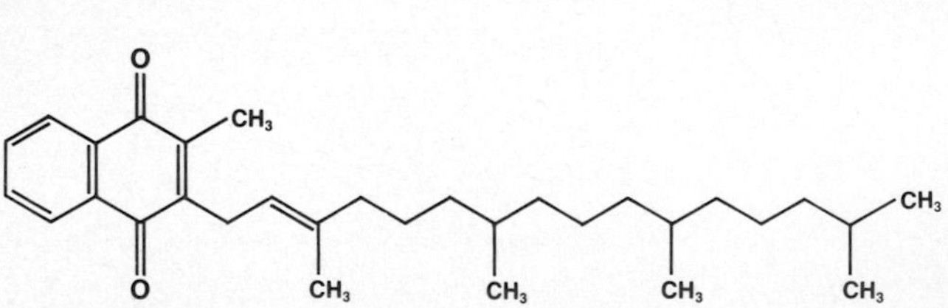

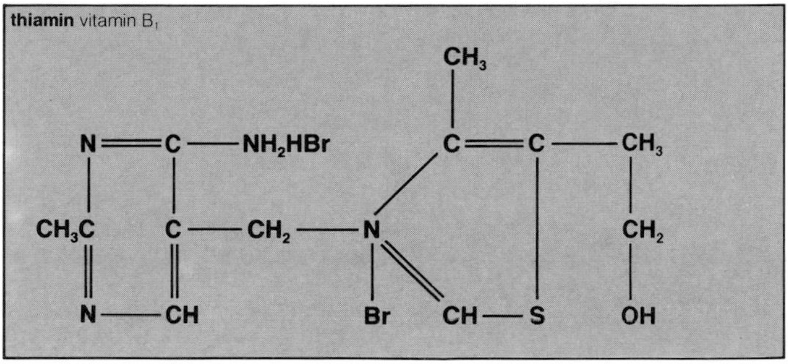

thiamin vitamin B_1

thiamin (*n*) vitamin B_1. Thiamin is a water soluble
(p.72) vitamin (p.72) found in the germ (p.108)
and bran (p.119) of all cereal grains (p.112). The
vitamin is also found in yeast extract (p.151) and
in meat (p.95), many vegetables, nuts and milk. A
deficiency (p.72) of thiamin leads to the disease
called beriberi (↓).

beriberi (*n*) a disease caused by lack of thiamin (↑)
in the diet (p.137) which leads to muscular (p.95)
weakness and deterioration (p.155) of the
nerves, leading to heart failure.

riboflavin (*n*) vitamin B_2. Riboflavin is a yellow
water soluble (p.72) vitamin (p.72) found in liver
(p.41), yeast extract (p.151), kidney (p.102),
cheese and eggs as well as many other foods.
Deficiency (p.72) of riboflavin results in soreness
of the eyes and mouth.

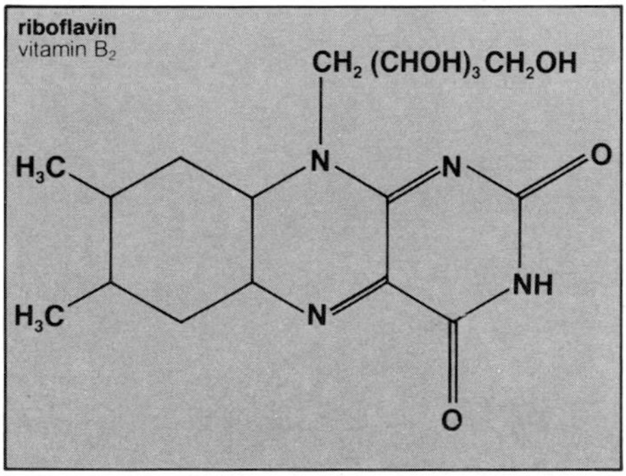

riboflavin vitamin B_2

vitamin B₆

CH₂OH

HO   CH₂OH   HO   CH₂OH   HO   CH₂OH

CHO   CH₂NH₂

H₃C   N   pyridoxine   H₃C   N   pyridoxal   H₃C   N   pyridoxamine

vitamin B₆ a group of three substances which are
derivatives (p.19) of 2-methylpyridine; these are
pyridoxine, pyridoxal and pyridoxamine. Vitamin
B₆ is found in many foods including meat (p.95),
liver (p.41), cereals (p.112) and pulses (p.109).
Deficiency (p.72) of vitamin B₆ is rare but can
cause fits in infants, anaemia (↓) and dermatitis.

pyridoxine (*n*) = vitamin B₆ (↑).

pyrodoxal (*n*) = vitamin B₆ (↑).

pyridoxamine (*n*) = vitamin B₆ (↑).

cobalamin (*n*) the general name for a group of
water soluble (p.72) organic (p.11) molecules
known as vitamin B₁₂. This vitamin (p.72) is found
only in animal tissues (p.32) such as liver (p.41),
milk, meat (p.95), fish and eggs. Deficiency (p.72)
of vitamin B₁₂ causes pernicious anaemia (↓).

cyanocobalamin (*n*) = cobalamin (↑).

pernicious anaemia a disease caused by a lack of
vitamin B₁₂ (↑) normally due to an inability to
absorb (p.41) the vitamin (p.72) in the gut (p.42).
Deficiency (p.72) of the vitamin in the diet (p.137)
can cause the disease; a particular problem for
vegans (p.143) who eat no animal products.

niacin (*n*) a water soluble (p.72) vitamin (p.72)
found in many plant and animal tissues (p.32)
including yeast (p.151), meat (p.95), liver (p.41),
fish, cheese, cereals (p.112) and pulses (p.109).
A deficiency (p.72) results in pellagra (↓).

nicotinic acid = niacin (↑).

pellagra (*n*) a disease caused by a lack of niacin
(↑). Because niacin is used in the body to obtain
energy from glucose (p.46) a lack of the vitamin
(p.72) affects tissues (p.32) which use a lot of
energy. Symptoms (p.143) include weakness in
the muscles (p.95), broken skin, loss of appetite,
tiredness and digestive (p.41) upsets.

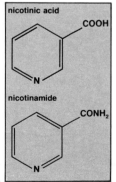

nicotinic acid

COOH

nicotinamide

CONH₂

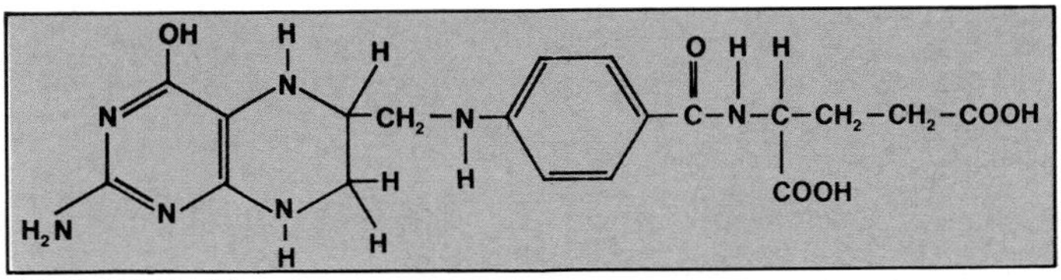

tetrahydrofolic acid

folic acid a group of water soluble (p.72) organic (p.11) compounds used in the synthesis (p.13) of nucleic acids (p.31). A deficiency (p.72) of folic acid leads to anaemia (↓).

anaemia (*n*) a disorder caused by a low number of red blood cells.

pantothenic acid

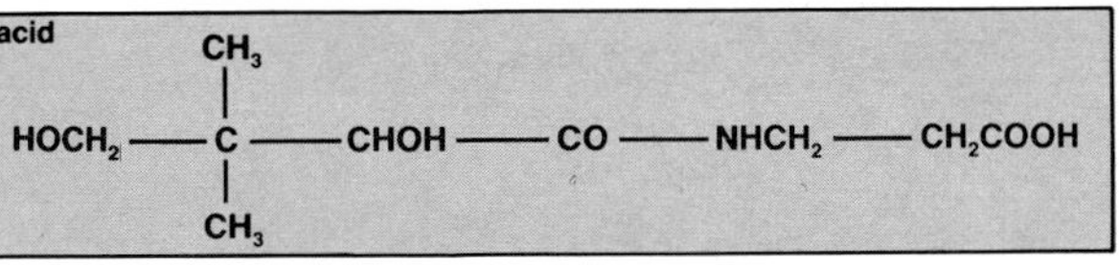

pantothenic acid an organic (p.11) molecule found in all living cells. Because it is so widespread in foodstuffs, deficiency (p.72) is not known. This vitamin forms part of co-enzyme A (p.34) and so is important in metabolism (p.144).

biotin (*n*) an organic (p.11) molecule found in many foods including egg yolk, cereals (p.112), kidney (p.102), liver (p.41), vegetables and yeast (p.151). It may not be needed in the diet (p.137) as it is synthesized (p.13) by intestinal bacteria (p.148).

biotin

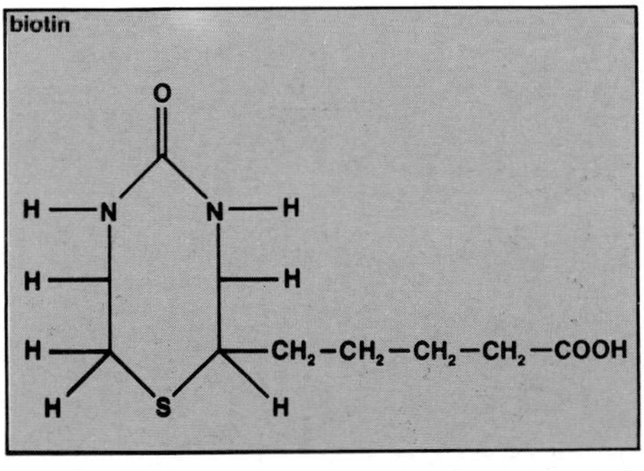

pigment (*n*) a molecule, compound or substance which is coloured. Most foods are coloured and therefore they contain pigments. These can be either natural colours or synthetic (p.13) compounds.

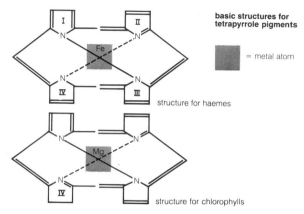

basic structures for tetrapyrrole pigments

= metal atom

structure for haemes

structure for chlorophylls

tetrapyrrole (*n*) a group of organic (p.11) molecules made up of a large ring of four smaller rings. The tetrapyrroles are found bonded (p.17) to different proteins (p.61) in plant and animal tissues (p.32), e.g. chlorophyll (p.30), haemoglobin (↓), myoglobin (↓). The large ring structure holds an atom of a metal element (p.8), usually magnesium (Mg) or iron (Fe), and it is this atom which gives the molecule its colour.

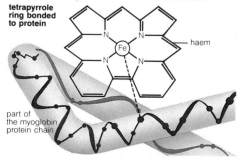

tetrapyrrole ring bonded to protein

haem

part of the myoglobin protein chain

haemoglobin (*n*) the red protein (p.61) in blood. It contains the iron-bearing tetrapyrrole (↑) haem (↓).

haem (*n*) a tetrapyrrole (↑) which contains an atom of iron (Fe). This iron can bond (p.17) to oxygen giving a bright red colour. In the absence of oxygen haem has a dull brown colour.

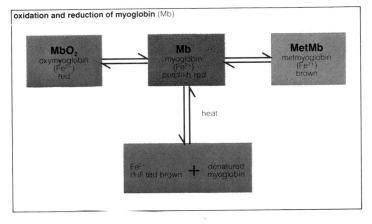

oxidation and reduction of myoglobin (Mb)

myoglobin (*n*) the red protein (p.61) of muscle (p.95). It contains haem (↑) and can bond (p.17) to oxygen and other atoms through the ferrous (Fe^{2+}) iron of the haem.

oxymyoglobin (*n*) the form of myoglobin (↑) containing oxygen bonded (p.17) to the ferrous (Fe^{2+}) iron of the haem (↑). Oxymyoglobin is bright red in colour.

metmyoglobin (*n*) myoglobin (↑) which does not contain oxygen and whose iron has been reduced (p.14) to the ferric (Fe^{3+}) form. Metmyoglobin is a grey-brown colour.

nitrosomyoglobin (*n*) myoglobin (↑) which has reacted (p.12) with nitrite (*see* nitrate p.20). Nitrosomyoglobin is a bright red colour. Nitrite is used in the curing (p.163) and preservation (p.89) of meats (p.95), particularly pig meat, and is responsible for the bright pink colour of such products through its reaction with myoglobin.

carotenoids (*n.pl.*) a group of red, yellow and
  orange pigments (p.78) found in most plants.
  They are not synthesized (p.13) by animals and
  so must be taken in the diet (p.137). Carotenoids
  are used by humans to make vitamin A.

β-carotene

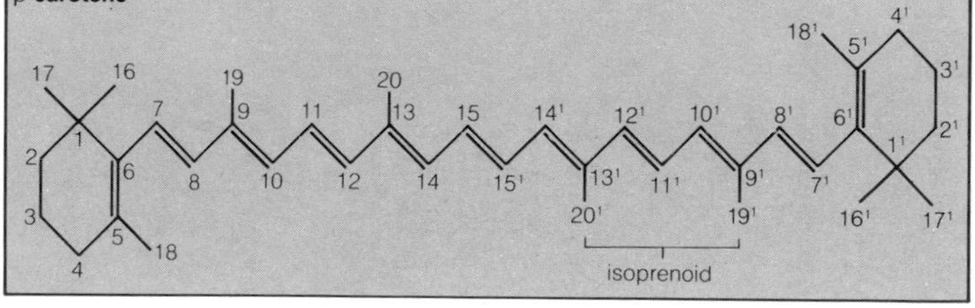

carotenes (*n.pl.*) the main pigments (p.78) of the
  carrot root (p.104). Carotenes are a kind of
  carotenoid (↑) found in many plant tissues (p.32)
  giving them a yellow or orange colour.

xanthophylls (*n.pl.*) a name sometimes given to
  carotenoids (↑) which contain hydroxyl (-OH)
  functional groups (p.19). Cryptoxanthin is an
  example of a xanthophyll.

β-ionone ring a partially unsaturated (p.18) ring
  found in carotenoids (↑). Carotenoids must have
  a β-ionone ring to be able to act as a vitamin A
  (*see* retinol (p.72)) precursor (or provitamin A).
  β-carotene has two β-ionone rings and therefore
  has twice the provitamin A activity of ∝ and
  -carotene (↑) which only have one β-ionone ring.

astaxanthin (*n*) an orange-pink carotenoid (↑)
  pigment (p.78) found in the muscle (p.95) of
  salmon and in many shellfish.

astaxanthin

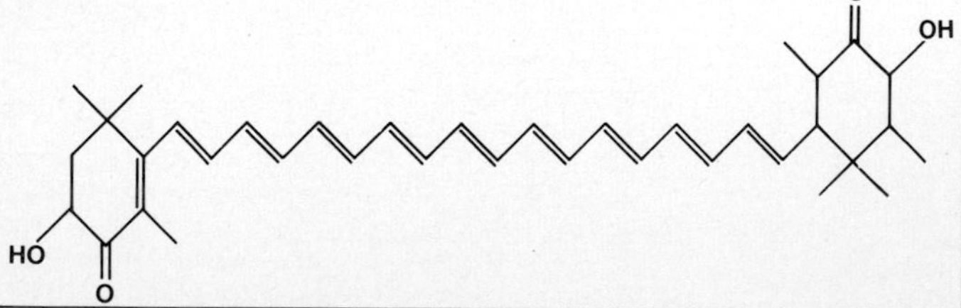

lycopene

lycopenes (*n.pl.*) a group of carotenoids (↑). The lycopenes give rise to red colours, e.g. in tomato and watermelon.

isoprenoids (*n.pl.*) the group of compounds which form the unsaturated (p.18) aliphatic (p.11) chains of the carotenoid (↑) structure. Isoprene is a five carbon (C_5) unit. Also known as **methylbuta-1,3-diene**.

colour changes in anthocyanins caused by changes in pH

anthocyanins (*n pl.*) a group of red glycoside (p.50) pigments (p.78) containing one or two carbohydrate (p.43) molecules and a set of aromatic (p.11) molecules. Anthocyanins are found in many fruit and vegetables, e.g. cherry, apple, red cabbage.

leucoanthocyanins (*n.pl.*) a group of colourless compounds found in many fruits and vegetables which may be hydrolysed (p.13) to a pink pigment (p.78). Hydrolysis of leucoanthocyanins causes a pink discolouration in damaged fruit of some kinds, e.g. pears, apples. The leucoanthocyanins are flavoured and can give fruit a sharp, dry taste.

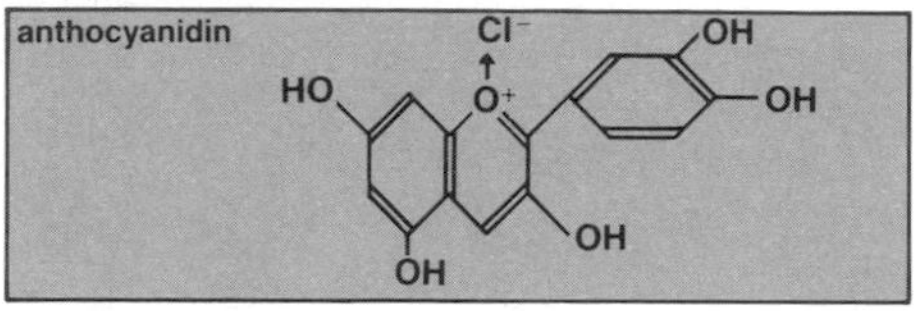

anthocyanidin

anthocyanidins (*n.pl.*) group of related chemicals
which contain a number of aromatic (p.11) rings
and hydroxyl (−OH) functional groups (p.19).
Anthocyanidins combine with sugars (p.44)
through the formation of glycosidic bonds (p.44)
to produce the coloured anthocyanins (p.81).

benzopyran derivatives a group of pigments
(p.78) based on aromatic (p.11) ring compounds
including anthocyanins (p.81) and flavones (↓).

flavones (*n.pl.*) group of benzopyran derivatives
(↑) related to the anthocyanins (p.81) which have
yellow or orange colours. Flavones are found in
onion skin, hops and tea.

flavonoids (*n.pl.*) a group of poorly-coloured
aromatic (p.11) glycosides (p.50) found in many
plants. They are not important as food colours
except in forming off-colours (↓) in some foods.

anthoxanthins (*n.pl.*) = flavonoids (↑).

off-colour (*n*) a pigment (p.78) produced in a food
which makes it look bad or unacceptable.

betalains (*n.pl.*) a group of red pigments (p.78)
found in the beetroot and other plants. They
consist of aromatic (p.11) and non-aromatic rings
and may contain carbohydrates (p.43).

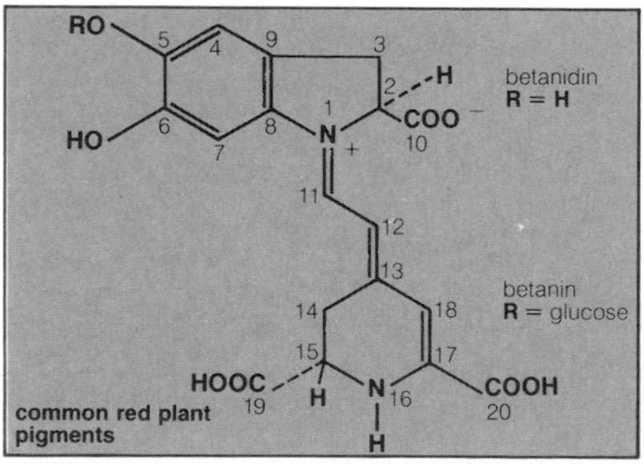

common red plant pigments

pheophytin (*n*) the compound formed from chlorophyll (p.30) when green plants are left to stand or heated. In pheophytin the magnesium (Mg) of the chlorophyll tetrapyrrole (p.78) is replaced by two atoms of hydrogen. Pheophytin is a pale grey-green colour.

melanoidins (*n.pl.*) the general name for a group of brown pigments (p.78) formed in foods by browning reactions (p.12). Melanoidins are formed from organic (p.11) compounds containing nitrogen (e.g. proteins (p.61), peptides (p.63) and amino acids (p.61)) and reducing sugars (p.47).

melanin (*n*) a brown pigment (p.78) found in the skin of humans which darkens on exposure to sunlight.

food dye a natural or synthetic (p.13) pigment (p.78) used to colour foods.

synthetic dye = food dye (↑).

annatto (*n*) a crude pigment (p.78) extracted (p.90) from the seed pods (p.109) of the plant *Bixa orellana*. Annatto is a mixture of coloured compounds of which bixin (↓) is one.

bixin (*n*) a carotenoid (p.80) pigment (p.78) found in annatto (↑). Bixin is the main pigment of annatto.

cochineal (*n*) a red pigment (p.78) extracted (p.90) from the dried bodies of *Coccus cacti*, a beetle found in Mexico and the Canary Islands. The main coloured compound in cochineal is carminic acid (↓).

carminic acid an aromatic (p.11) compound containing three rings of carbon atoms found in cochineal (↑). When carminic acid reacts (p.12) with aluminium (Al) it forms a lake (↓) called carmine.

lake (*n*) an insoluble (p.23) pigment (p.78) which is produced by reacting (p.12) a food dye (↑) with aluminium (Al) or calcium (Ca). Lakes of water soluble (p.23) dyes are used for colouring aqueous (p.23) and non-aqueous foods.

turmeric (*n*) the powdered root of the plant *Curcuma longa* with a bright yellow-orange colour.

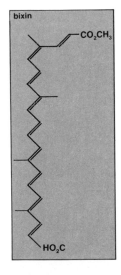

bixin
CO₂CH₃
HO₂C

additive (*n*) a substance added to food to aid
   processing (p.87), preservation (p.89), to
   improve flavour or colour, e.g. pigments (p.78),
   lakes (p.83), emulsifiers (↓), flavouring,
   humectants (↓), stabilizers (↓), vitamins (p.72),
   minerals (p.16). *See also* appendix five p.174.

emulsifier (*n*) a substance which allows the mixing
   of two or more immiscible (p.27) liquids (p.22) to
   form a stable (p.167) emulsion (p.26). Emulsifiers
   work by coating the surface of droplets of one
   liquid in such a way that they can stay dispersed
   (p.23) in the second liquid. **emulsify** (*v*).

emulsifying agent = emulsifier (↑).

surfactant (*n*) = emulsifier (↑). Surfactants are also
   used to coat the surface of a solution to stop
   evaporation (p.24).

anti-foaming agent a substance which stops the
   formation of foam (p.26) in a food.

humectant (*n*) a substance which absorbs (p.41)
   water and is used to keep foods moist, e.g.
   glycerol (p.45), invert sugar (p.47) and glucose
   (p.46) syrup (p.50).

stabilizer (*n*) a substance which allows food
   compounds which do not mix well to be mixed
   and stay in a homogeneous (p.24) state.

food colour *see* food dye (p.83).

thickener (*n*) an additive (↑) which makes a food
   more viscous (p.23).

binding agent an additive (↑) which holds a food
   together by sticking particles (p.24) and
   aggregates (p.25) together.

adhesive (*n*) = binding agent (↑).

clarifying agent an additive (↑) used in the
   production of beverages (p.93), to obtain a clear,
   bright solution (p.22), e.g. pectinase (p.50).

hydrocolloid (*n*) one of a group of natural or
   synthetic (p.13) polysaccharides (p.44) which
   bind water and form gels (p.25). Hydrocolloids
   are also called gums (↓) and are added to many
   foods as extenders (↓), stabilizers (↑), emulsifiers
   (↑) and adhesives (↑).

extender (*n*) an additive (↑) which increases the
   bulk of a food without necessarily adding to its
   nutrient (p.137) content.

gum (*n*) = hydrocolloid (↑).

agar, alginates,
monosodium glutamate
products of seaweed
by extraction and
purification

agar (*n*) a polysaccharide (p.44) hydrocolloid (↑) extracted (p.90) from seaweed. As well as being added to foods, agar is used for growing micro-organisms (p.147) in microbiological (p.147) tests.

alginates (*n.pl.*) a group of polysaccharide (p.44) hydrocolloids (↑) extracted (p.90) from seaweed.

carrageenan (*n*) a mixture of two polysaccharide (p.44) hydrocolloids (↑) which contain sulphate (p.20) and can be extracted (p.90) from seaweed.

gum arabic an exudate (↓) obtained from the acacia tree. It is a hydrocolloid (↑) which is widely used in the food industry to improve the texture (p.27) and viscosity (p.27) of foods.

tragacanth (*n*) an exudate (↑) gum (↑) obtained from the shrub-like plants of the *Astragalus* family.

exudate (*n*) a substance given off by a plant or animal.

monosodium glutamate the sodium salt (p.16) of the amino acid (p.61) glutamic acid. Monosodium glutamate (MSG) is added to many foods as a flavour enhancer (↓). Regularly eating foods containing MSG (used in commercial Chinese cooking) can give rise to an illness known as Chinese restaurant syndrome with symptoms (p.143) of headaches, dizziness, muscle cramps and nausea, amongst others.

flavour enhancer a compound or substance added to foods which brings out or increases flavours already present in that food.

enhancer (*n*) see flavour enhancer (↑).

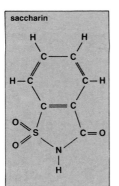

saccharin

potentiate (*v*) to make more powerful or give power. Enhancers (↑) are said to potentiate flavours in foods by making the natural flavour more powerful. **potentiator** (*n*).

flavour modifier a chemical additive which can reduce or enhance (↑) the natural flavour of a food. Flavour modifiers do not add flavour to a food themselves.

sweetener (*n*) any compound used to make foods sweet.

saccharin (*n*) benzoic sulphimide. A chemical sweetener (↑) added to foods in place of sucrose (p.46). Saccharin is 550 times sweeter than sucrose.

cyclamate (*n*) sodium cyclo-hexyl-sulphamate. A chemical sweetener (p.85) 30 times sweeter than sucrose (p.46). Cyclamate is now banned for use in foods in many countries because it is known to be a possible carcinogen (p.142).

anti-spattering agents compounds which are added to fats (p.52) and oils (p.52) to stop the condensation (p.24) of water drops in lipids (p.51).

anti-staling agent a chemical which slows down the process of staling (p.121) in baked products and helps to keep the crumb (p.121) soft, e.g. sucrose (p.46).

anti-caking agent a chemical additive (p.84) which can absorb (p.41) relatively high amounts of moisture without becoming wet. Anti-caking agents are added to powders (like table salt) and other dried foods to keep them dry, stop them forming lumps and allow them to remain free flowing. Examples include calcium phosphates, magnesium oxide and some fatty acids (p.53).

firming agent a chemical added to vegetable products to prevent loss of turgor (p.111) and therefore textural (p.27) properties. Firming agents work by preventing the loss of moisture from the plant tissue. Common salt can act as a firming agent but calcium compounds are more common.

crisping agent = firming agent (↑).

sequestrant (*n*) a chemical which is able to bind (p.84) other chemicals, notably metal ions (p.9). Sequestrants are also known as chelators (↓). **sequestrate** (*v*), **sequestrated** (*adj*).

chelator (*n*) a chemical which can bind (p.84) metal ions (p.9) very strongly, e.g. ethylene diamine tetraacetic acid (EDTA) which is synthetic (p.13), and citric acid (p.21) and phosphoric acid. Chelators are very useful food additives (p.84). By binding metal ions, which often cause deterioration (p.155) in foods by catalyzing (p.68) various chemical reactions (p.12), chelators prevent the unwanted action of metal ions. **chelation** (*n*), **chelate** (*v*).

struvite (*n*) the fine, sharp crystals (p.11) found in some canned fish products. The formation of struvite can be inhibited (p.70) by the addition of a chelator (↑) or sequestrant (↑).

process (*v*) to treat a food in such a way as to change its nature and properties in order to preserve (p.155) it, to improve its eating quality or to make useful ingredients, e.g. to extract (p.90) juice (p.90) from fruit, to make meat (p.95) products from comminuted meat (p.103). There are several ways in which food is processed: (1) cooking (↓); (2) blanching (p.89); (3) methods of preservation (p.89), such as drying (p.162), freezing (p.159), chilling (p.158), canning (p.156); (4) extraction; (5) refinement (p.90); fabrication (p.90). **process** (*n*).

raw material a natural product from which a food can be prepared. Raw materials can either be already edible (p.111), e.g. fruits, or may need processing before they can be eaten, e.g. meats (p.95), coffee (p.94) beans.

thermal processing any process using heat.

cooking (*n*) a process where food is made more pleasant to eat and more digestible (p.136) by causing, e.g. a breakdown of connective tissue (p.97) in animal tissues (p.32) and cellulose (p.49) in plants. It also causes the destruction of bacteria (p.148), moulds (p.151) and other contaminants (p.152), e.g. most effectively, sterilization (p.151) and pasteurization (p.155). Cooking causes some loss of nutrients (p.137), e.g. water-soluble (p.23) minerals (p.16) and vitamins (p.72) especially vitamin C (p.74) and thiamin (p.75). It also causes structural changes in foods, e.g. coagulation (p.67) of proteins (p.61), breakdown of polysaccharides (p.44). **cook** (*v*).

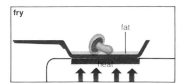
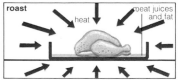

fry (*v*) to cook (↑) in a pan with fat (p. 52).
roast (*v*) to cook (↑) in a closed oven with dry heat (p.88). In roasting meat (p.95), juices (p.90) are forced out and evaporate (p.24) on the surface to produce browning. *See* Maillard browning (p.92).

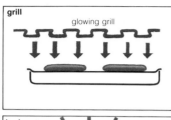

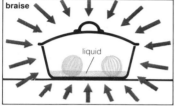

dry heat direct heat applied to food not suspended
(p.23) in aqueous (p.23) solutions (p.22), water,
fat (p.52) or oil (p.52).

grill (*v*) to cook (p.87) using radiant heat.

stew (*v*) to cook (p.87) in a closed container at a
low temperature.

braise (*v*) to cook (p.87) in a closed container with
a small amount of liquid, in the oven (p.168). the
evaporated (p.24) water, i.e. steam, helps the
cooking process.

boil (*v*) to cook (p.87) in water at its boiling point
(p.25), i.e. 100°C.

microwave cooking the use of high energy
electromagnetic radiation (↓) to cook (p.87)
foods. **microwave** (*n*), **microwave** (*v*).

electromagnetic radiation energy in wave form
such as light, ultra-violet light, X-rays and
microwaves (↑). All electromagnetic radiation
travels at the same speed, i.e. 3×10^8 ms^{-1}.
When electromagnetic radiation is passed
through a substance it may be reflected, be
absorbed (p.41) or pass straight through. As it is
energy, if it is absorbed it passes energy to the
molecules of the substance which absorbs it. It is
in this way that food heats up when microwaves
are passed through it.

blanch (*n*) to part cook (p.87) by dropping into hot water, (82–93°C), for 1–5 minutes. Fruit and vegetables are blanched before canning (p.156), drying (p.162), or freezing (p.159) to soften their texture (p.27), destroy enzymes (p.68), remove unpleasant flavours and air, or cause shrinkage (↓). The destruction of enzymes in plant foods by denaturation (p.66) through heating is important before freezing since without their destruction, they would continue to catalyse (p.68) unwanted reactions at a slow rate and lower the quality (p.128) of the food. Blanching can be used to remove unwanted salt (p.16) or skins. It can cause the loss of 10–20% of sugars (p.44), salts, protein (p.61), niacin (p.76), some of the B vitamins (pp 75, 76) and up to one third of the vitamin C (p.74). **blanching** (*n*).

steam blanching a method of blanching (↑) foods in steam. It can be very useful in preventing losses of nutrients (p.137) during blanching, e.g. loss of water-soluble (p.72) vitamins (p.72) through leaching (p.72).

shrinkage (*n*) the way in which food becomes smaller during storage or processing.

preservation (*n*) any process used to allow food to be stored for longer before use. **preserve** (*v*).

case hardening the process by which small molecules, such as salts (p.16) and sugars (p.44), gather at the surface of foods during dehydration (p.162) and form a skin.

peel (*v*) to remove the outer skin, rind, covering or husk. Particularly used of fruits. **peel** (*n*), **peeling** (*n*).

case hardening

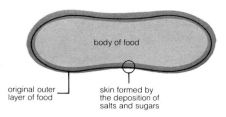

body of food

original outer
layer of food

skin formed by
the deposition of
salts and sugars

juice extraction

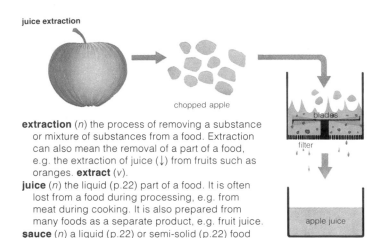

chopped apple

blades

filter

apple juice

extraction (*n*) the process of removing a substance or mixture of substances from a food. Extraction can also mean the removal of a part of a food, e.g. the extraction of juice (↓) from fruits such as oranges. **extract** (*v*).

juice (*n*) the liquid (p.22) part of a food. It is often lost from a food during processing, e.g. from meat during cooking. It is also prepared from many foods as a separate product, e.g. fruit juice.

sauce (*n*) a liquid (p.22) or semi-solid (p.22) food made from different products and served with another food.

refine (*v*) to purify a food or part of a food product, e.g. the production of sugar (p.44). **refinement** (*n*), **refined** (*adj*).

fabrication (*n*) a method by which foods are made from processed natural or synthetic (p.13) materials. **fabricate** (*v*).

fabricated food food made from synthetic (p.13) or processed natural products, e.g. food analogues (↓), foods such as meat (p.95) pies, cakes, other baked foods, pasta (p.112) and breakfast cereals.

fabricated food

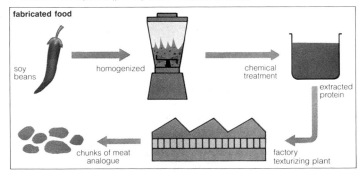

soy beans   homogenized   chemical treatment

extracted protein

chunks of meat analogue   factory texturizing plant

food analogue a food made to appear, smell, feel and taste the same as a natural food. Food analogues are fabricated foods (↑), e.g. margarine (p.60), meat (p.95) analogue made from textured vegetable protein (p.67).

engineered food = fabricated food (↑).

fast food a general term used for foods where most of the preparation has been done before service. Most fast foods are prepared for eating by frying (p.87) or microwave (p.88).

convenience food any food prepared by industrial processes so that it may be easily cooked (p.87) and served or served cold, e.g. canned (p.156) foods, frozen (p 159) foods, dried foods, ready-made food sauces (↑), cook-chill (p.158) foods.

instant food dried foods which reconstitute (↓) quickly when water is added, e.g. instant coffee, instant tea, instant potatoes (↓), instant milk (↓).

reconstitution (*n*) a process by which instant foods (↑) or foods stored in some other way may be returned to their natural state, e.g. by adding water. **reconstitute** (*v*).

instant potato dried, powdered potato which can be quickly reconstituted (↑) with water.

instant milk dried, powdered milk which may be reconstituted (↑) by the addition of water.

food reactions chemical reactions (p.12) in foods brought about by storage or processing.

enzymic browing a number of processes in which oxidoreductase (p.69) enzymes (p.68) present in plant tissues (p.32), e.g. phenolases (↓), catalyse (p.68) the formation of brown pigments (p 78). The substrates (p.70) for these enzymes are natural compounds in the food and oxygen. The process usually spoils the food and reduces its eye appeal (p.130). Enzymic browning can be inhibited (p.70) by cooking (p.87) or blanching (p.89) the food so that the enzymes are denatured (p.66) and are no longer active or by removing oxygen, e.g. by vacuum (p.22) packing, or adding chemicals which react (p.12) with oxygen.

phenolases (*n.pl.*) a group of oxidoreductases (p.69) found in plant tissues (p.32) which cause enzymic browning (↑).

catechol (*n*) a phenolic compound (p.11) consisting of a benzene ring and two hydroxyl (–OH) functional groups (p.19) commonly found in fruit, particularly apples. Catechol is an important example of the diphenolic substrates (p.70) of the polyphenolases (p.91) which cause enzymic browning (p.91).

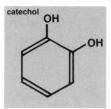

catechol

non-enzymic browning a group of reactions (p.12) between sugar (p.44) molecules or between sugars and proteins (p.61) or amino acids (p.61), or the breakdown of ascorbic acid, which produces a brown colouring in foods either on heating or storage. There are three kinds of non-enzymic browning: (1) Maillard browning (↓); (2) caramelization (↓); (3) ascorbic acid browning (↓).

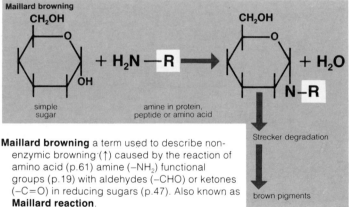

Maillard browning

simple sugar

amine in protein, peptide or amino acid

Strecker degradation

brown pigments

Maillard browning a term used to describe non-enzymic browning (↑) caused by the reaction of amino acid (p.61) amine (–NH$_2$) functional groups (p.19) with aldehydes (–CHO) or ketones (–C=O) in reducing sugars (p.47). Also known as **Maillard reaction**.

sugar amine the first reaction (p.12) product of Maillard browning (↑) when a reducing sugar (p.47) reacts with an amino acid (p.61).

Strecker degradation a reaction (p.12) in the process of Maillard browning (↑) which involves breakdown of the amino acid (p.61) after it has reacted with the reducing sugar (p.47).

caramelization (*n*) the reaction (p.12) of sugars (p.44) with each other when they are heated, especially when there is no water present. Caramelization results in the formation of brown substances with a characteristic flavour.

alcholic beverage
fermentation

fruit juice (grape) or
water extract of cereal
grain

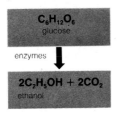

$C_6H_{12}O_6$
glucose

enzymes

$2C_2H_5OH + 2CO_2$
ethanol

isosacchrosan (*n*) a product of the sucrose
(p.46) molecule made by the loss of one
molecule of water. Isosacchrosan is formed
during the first stages of caramelization (↑) and
is not sweet.

ascorbic acid browning the process by which
ascorbic acid (p.74) breaks down to give brown-
coloured products. This reaction (p.12) can
happen under both aerobic (p.32) and anaerobic
(p.32) conditions.

brewing (*n*) a process by which alcoholic
beverages (↓) are made. The word brewing is
normally used of beers (↓).

beverage (*n*) a drink.

alcoholic beverage (*n*) any drink containing
alcohol, i.e. ethanol (p.21).

beer (*n*) an alcoholic beverage (↑) made by the
fermentation (p.34) of malted barley (p.114).

cider (*n*) an alcoholic beverage (↑) made by the
fermentation (p.34) of apple juice (p.90).

wine (*n*) an alcoholic beverage (↑) made by the
fermentation (p.34) of grape juice (p.90) or the
juices of other fruits.

vinification (*n*) the process by which wine (↑) is
made.

spirits the group of alcoholic beverages (↑) which
contain large amounts of ethanol (p.21), e.g. gin,
whisky, brandy, vodka, saki. Spirits are made by
concentrating (p.22) the ethanol by distilling (↓) it
out of a fermented (p.34) liquid (p.22).

distillation

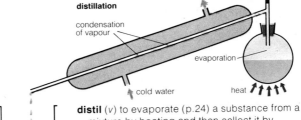

condensation
of vapour

evaporation

cold water   heat

distillate

distil (*v*) to evaporate (p.24) a substance from a
mixture by heating and then collect it by
condensation (p.24), e.g. the distillation of
ethanol (p.21) to make spirits (↑) **distillation** (*n*).

lager (*n*) *see* beer (↑).

liqueur (*n*) *see* spirits (↑).

coffee (*n*) a non-alcoholic beverage (p.93) produced from the dissolution (p.22) and suspension (p.23) in hot water of up to 200 different compounds from the roasted, milled beans of the coffee plant. One of the more important chemicals in coffee is caffeine (↓).

coffee cherry a bright red fruit of the coffee tree from which coffee (↑) is made.

coffee bean the seed (p.107) contained in the red fruit, the coffee cherry (↑), of the coffee tree. There are two coffee beans in each coffee cherry.

tea (*n*) a non-alcoholic beverage (p.93) made from an infusion of the dried leaves of the tea bush. Tea contains a variety of compounds but tannins and caffeine (↓) are the best known.

caffeine (*n*) a simple organic (p.11) compound consisting of two partially unsaturated (p.18), nitrogen-containing rings and methyl (−CH₃) functional groups (p.19). Caffeine is a stimulant.

stimulant (*n*) a compound which increases the activity of bodily functions or excites emotions.

cocoa bean the seed (p.107) of the cocoa plant. They grow in pods (p.109) and form the raw material from which chocolate (p.127) is made. The beans are removed from the pods after fermentation (p.34) which develops the flavour and colour, and chocolate, a plant fat (p.52), is extracted.

clarification (*n*) a process which removes very small particles (p.24) in colloidal (p.25) dispersion (p.23) from liquids. Clarification makes liquids clear and bright. The particles can be removed by filtration (p.24) or the use of a chemical to form large enough particles to settle out of the liquid, e.g. isinglass (↓). **clarify** (*v*).

isinglass (*n*) gelatin (p.98) used to clarify (↑) alcoholic beverages (p.93). Isinglass is usually made from fish swim bladders.

contamination (*n*) a process by which food is made inedible (p.111), e.g. by the growth of food-poisoning (p.152) bacteria (p.148), moulds (p.151) or yeasts (p.151), from lack of cleanliness or already contaminated food. **contaminate** (*v*).

carton (*n*) a container, usually made from plastic or cardboard, in which foods are stored, e.g. milk.

coffee cherry

bean

pulp

outer skin (red)

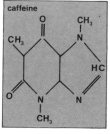

caffeine

carton

muscle (*n*) a part of the body of an organism which produces movement by contraction (p.97). Muscle contains large amounts of special proteins (p.61) which can contract, as well as some carbohydrate (p.43), lipid (p.51) and other compounds.

meat (*n*) the muscle (↑) of any animal, e.g. beef (p.102), lamb (p.101), pork (p.102), mutton (p.101), veal (p.101), chicken, other poultry (p.101), which is obtained after the death of the animal. Meat generally contains 20% protein (p.61), 10–30% lipid (p.51), some carbohydrate (p.43) and the remainder is water.

fibre of voluntary muscle

striped band

skeletal muscle muscle (↑) joined to bone and used in the body of an animal for movement. Skeletal muscle is made up of muscle fibres (↓) which have a striped appearance. It is known as striped muscle or striated muscle. Because these muscles can be controlled by conscious thought they are also called voluntary muscles. Skeletal muscle consists of bundles of long, tube-like cells which are multinucleate (p.30) and are called muscle fibres.

striped muscle = skeletal muscle (↑).

striated muscle = skeletal muscle (↑).

voluntary muscle = skeletal muscle (↑).

structure of striated muscle

endomysium
nucleus
A band
I band
H-line
Z-line
myofibril

smooth muscle muscle (↑) which has no striped appearance and is not joined to bone. Because smooth muscle cannot be controlled by the brain through conscious thought it is also called involuntary muscle. Smooth muscle consists of small cells each with its own nucleus (p.30).

involuntary muscle = smooth muscle (↑).

unstriated muscle = smooth muscle (↑).

muscle fibre the long, thin multinucleate (p.30) cells of muscle (↑) tissue (p.32). Muscle fibres contain two main kinds of contractile protein (p.62), myosin (p.96) and actin (p.96). Each muscle fibre is surrounded by a cell membrane (p.32) and a tube of connective tissue (p.97) called the endomysium (p.97).

myofibre (*n*) = muscle fibre (↑).

fibre bundle a group of muscle fibres (↑) held together by connective tissue (p.97) called the perimysium (p.97).

myofilament (*n*) the long, string-like structures within muscle fibres (p.95) made mainly of the muscle proteins (p.61) myosin (↓) and actin (↓). About 2000 myofilaments are found in each muscle fibre. Myofilaments are made up of repeating units called sarcomeres (↓).

myofibril (*n*) = myofilament (↑).

thin filament the long, string-like structure made of actin (↓) and other minor muscle (p.95) proteins (p.61). The thin filaments combine with the thick filaments (↓) to make the myofilaments (↑).

thick filament the long, string-like structure made of myosin (↓) and other minor muscle (p.95) proteins (p.61). The thick filaments combine with the thin filaments (↑) to make the myofilaments (↑).

Z-line (*n*) an area in the myofilament (↑) which seems to hold the thin filaments (↑) together.

sarcomere (*n*) the distance between two Z-lines (↑) in the myofilament (↑).

sliding filament hypothesis

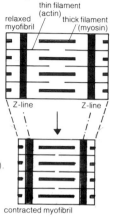

contracted myofibril

actin arranged in double-helical structure

actin (*n*) a globular protein (p.61) which takes part in the contraction (↓) of muscle (p.95) through the formation of the thin filaments (↑) and their interaction (p.16) with the thick filaments (↑).

myosin (*n*) a fibrous protein (p.61) which takes part in the contraction (↓) of muscle (p.95) through the formation of thick filaments (↑) and their interaction (p.16) with thin filaments (↑).

tropomyosin B (*n*) a minor protein (p.61) of muscle fibres (↑) which is found bonded (p.17) to the thin filaments (↑).

troponin (*n*) a minor protein (p.61) of myofilaments (↑).

actomyosin (*n*) a protein (p.61) aggregate (p.25) formed when actin (↑) and myosin (↑) interact (p.16) during muscle contraction (↓).

myosin

myosin head

heavy meromyosin

light meromyosin

two polypeptide chains in α-helical supercoil

contraction (*n*) the process by which an object becomes smaller. In muscle (p.95), contraction is caused by the interaction (p.16) of actin (↑) in the thin filaments (↑) with myosin (↑) in the thick filaments (↑) to form actomyosin (↑). The opposite of contraction is relaxation. **contract** (*v*).

sarcolemma (*n*) the cell membrane (p.32) of muscle fibres (p.95).

connective tissue any tissue (p.32) in the body of an animal which acts as a support or joins other tissues or cells together, e.g. skin, bone, tendon. Connective tissues consist mainly of the fibrous protein (p.61) collagen (p.98) but also may contain other proteins like elastin (p.98), and polysaccharides (p.44). The connective tissue in muscle (p.95) is the yellow-white tissues surrounding the muscle fibres (p.95), between the fibre bundles (p.95) and surrounding the whole muscles. These tissues are called the endomysium (↓), perimysium (↓) and epimysium (↓) respectively.

endomysium (*n*) the connective tissue (↑) surrounding each muscle fibre (p.95).

perimysium (*n*) the connective tissue (↑) between the fibre bundles (p.95) which forms part of the marbling (p.98) of meat (p.95).

epimysium (*n*) the connective tissue (↑) surrounding each muscle (p.95).

tendon (*n*) the thick rope of connective tissue (↑) made mainly of collagen (p.98) joining muscle (p.95) tissue (p.32) to bone in skeletal muscles (p.95). Tendons are inelastic and do not stretch.

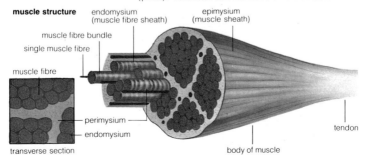

muscle structure

endomysium (muscle fibre sheath)

epimysium (muscle sheath)

muscle fibre bundle

single muscle fibre

muscle fibre

perimysium

endomysium

transverse section

body of muscle

tendon

ligament (*n*) the elastic connective tissue (p.97) which joins bones to other bones particularly at joints (↓). Because ligaments stretch and return to their original length after stretching, they help to control the movement of joints.

joint[1] (*n*) a region in the body where two or more bones come into contact, e.g. the human knee and elbow joints.

elastin (*n*) a protein (p.61) of the connective tissue (p.97) found in muscle (p.95).

ground substance a material rich in protein (p.61) and carbohydrate (p.43) found in the connective tissue (p.97).

collagen (*n*) the main fibrous (p.65) protein (p.61) of the connective tissues (p.97) which is important in determining the texture (p.27) of meat (p.95).

gelatin (*n*) denatured (p.66) or partly broken down collagen (↑). Gelatin is extracted (p.90) from animal bone and skin and used in the food industry in many products, e.g. jelly, as a binding agent (p.84), as isinglass (p.94).

marbling (*n*) the net-work of white connective tissue (p.97), i.e. perimysium (p.97), and fat (p.52) seen in muscle tissue (p.32).

slaughter (*n*) the process of killing animals.

carcass (*n*) the body of a slaughtered (↑) animal: also carcase.

butchery (*n*) the process of cutting up the carcass (↑) into pieces for cooking (p.87).

hot de-boning a process in which all of the muscle (p.95) tissue (p.32) is removed from the bones soon after slaughter (↑) of the animal.

primal cut a large portion of the carcass (↑) cut out for conditioning (↓). Each primal cut contains several different pieces to be butchered (↑) out after conditioning.

cold shortening the contraction (p.97) of muscle (p.95) caused by too rapid cooling after slaughter (↑). Cold shortening can give very tough meat (p.95).

drip (*n*) the process by which liquid (p.22) is lost from meat (p.95) during storage after butchery (↑). Also used as the name of the liquid lost from the meat.

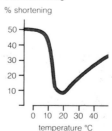

cold shortening

% shortening

temperature °C

water-holding-capacity the ability of proteins (p.61) and muscle (p.95) proteins in particular to bind water.

conditioning (*n*) the process in which meat (p.95) is stored at a low temperature (0–4°C) for several days before cooking (p.87). Conditioning is also known as *aging* and *maturing* and makes the meat more tender (↓). Enzymes (p.68) in the meat breakdown the structure of stiff actomyosin (p.96) formed during rigor mortis (↓) so that the meat becomes softer.

post-mortem after death.

rigor mortis the stiffening of muscle (p.95) after death which makes the meat (p.95) tough and inedible (p.111).

resolution of rigor the process by which meat (p.95) or game (p.101) is softened during conditioning (↑). Proteolytic enzymes (p.69) including calcium activated neutral proteinase (↓) and the cathepsins (↓) present in the meat breakdown the structure of the contracted (p.97) muscle proteins thereby tenderizing (↓) it.

calcium activated neutral proteinase CANP. A proteolytic enzyme (p.69) found in meat (p.95) which breaks down the contracted (p.97) muscle proteins (p.61) during conditioning (↑).

cathepsin (*n*) one of a group of proteolytic enzymes (p.69) responsible for the breakdown of proteins (p.61) during conditioning (↑).

ultimate pH the pH (p.14) of meat (p.95) reached after rigor mortis (↑). Normally about 5.3, this pH is near to the optimum pH (p.71) of the cathepsins (↑). **pH$_u$** (*abbr*).

tenderizing (*n*) a process by which tough meat (p.95) is made tender (less tough). **tenderize** (*v*).

mechanical tenderizing the use of machines to break up the structure of tough meat (p.95) in order to make it tender.

enzyme tenderizing the use of proteolytic enzymes (p.69) to tenderize (↑) tough meat (p.95) by breaking down the structure of meat proteins (p.61). Arterial injection (p.100) and multiple-needle injection (p.100) are two methods of using enzymes such as papain (p.100), bromelain (p.100) and ficin (p.100).

papain (*n*) a proteolytic enzyme (p.69), from papaya (p.111) or paw-paw juice (p.90), used in the tenderizing (p.99) of meat (p.95), sometimes called vegetable pepsin (p.38). Its reaction (p.12) increases at high temperature and so continues to tenderize meat at the beginning of cooking (p.87). It is extremely active and readily degrades most proteins to very small fragments.

ficin (*n*) a proteolytic enzyme (p.69) of the fig used in tenderizing (p.99) meat (p.95). It is similar to papain (↑).

bromelain (*n*) a proteolytic enzyme (p.69) of the pineapple used to tenderize (p.99) meat (p.95). It is similar in action to papain (↑).

arterial injection a method of getting tenderizing (p.99) enzymes (p.68) into the blood of animals before slaughter (p.98). The enzyme becomes spread throughout the body to all muscles (p.95).

multiple-needle injection a method of getting tenderizing (p.99) enzymes (p.68) into meat (p.95) after slaughter (p.98).

tumbling (*n*) a method of rolling meat (p.95) inside a machine to bring about mechanical tenderizing (p.99). Used mainly with cured (p.163) meats.

massaging (*n*) = tumbling (↑).

tumbler massager

paddles

massage cabinet

multiple-needle injection

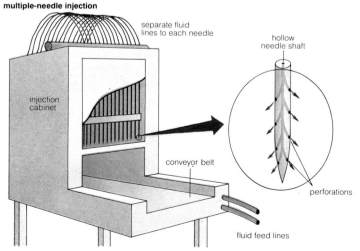

separate fluid lines to each needle

hollow needle shaft

injection cabinet

conveyor belt

perforations

fluid feed lines

polyphosphate (*n*) phosphates (p.20) of various kinds added to meat (p.95) by multiple-needle injection (↑) or by tumbling (↑) and massaging to increase water uptake and the water-holding capacity (p.99). Polyphosphates can also stop discolouration and speed the entry of brine (p.163) into meat during curing (p.163).

marinade (*n*) a mixture of wine (p.93), or vinegar (*see* ethanoic acid (p.21)), oil (p.52), and lemon juice (p.90) in which meat (p.95) is left to flavour and tenderize (p.99). **marinate** (*v*).

game (*n*) a wild animal or bird.

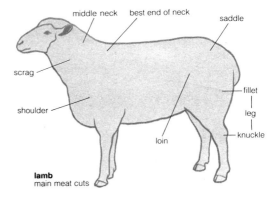

middle neck · best end of neck · saddle · scrag · fillet · shoulder · leg · knuckle · loin

lamb
main meat cuts

lamb (*n*) a young sheep up to 12 months or a spring lamb, from 3–6 months old. Also meat (p.95) from the lamb.

mutton (*n*) the meat (p.95) of sheep older than 1 year.

veal (*n*) the meat (p.95) of a calf (↓) fed on milk.

bull (*n*) male cattle.

bull calf young male cattle.

cow (*n*) female cattle.

calf (*n*) young of cattle.

steer (*n*) a castrated bull (↑).

heifer (*n*) a young cow (↑).

poultry (*n*) the meat (p.95) of domestic birds, e.g. chicken, turkey.

offal (*n*) the parts of the dead animal that are cut away and are not classed as muscle (p.95), e.g. brain, kidneys (↓), liver (p.41), tripe (↓) and chitterlings (↓).

kidney (*n*) one of a pair of organs (p.40) which remove excess water and unwanted compounds from the blood as it circulates through them. Urea is an important waste compound from amino acids (p.61) which is removed from the body in this way.

tripe (*n*) the lining of the stomach (p.37) of ruminants, e.g. calf (p.101).

chitterlings (*n*) the intestines of calf (p.101), steer (p.101) or pig.

beef
main meat cuts

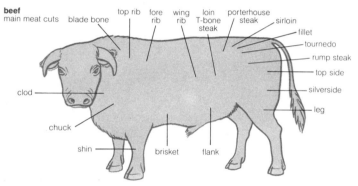

beef (*n*) the meat (p.95) of cattle.

ham (*n*) the back leg of a pig. Ham is also the name given to cured (p.163) pig meat (p.95).

pork (*n*) the fresh meat (p.95) from a pig which has not been salted or cured (p.163).

bacon (*n*) pig meat (p.95) produced by curing (p.163) the carcass (p.98) or portions of meat in salt solutions (p.22) called brine (p.163). The brine normally contains 25–30% by weight of sodium chloride, 2.5–4% potassium nitrate and often salt-tolerant bacteria (p.148) which convert potassium nitrate to potassium nitrite. The nitrite is broken down to nitric oxide which combines with myoglobin (p.79) to give the pink colour of bacon. Curing can take several days.

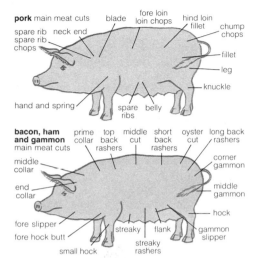

pork main meat cuts · blade · fore loin · loin chops · hind loin · fillet · chump chops

spare rib · neck end
spare rib
chops

fillet

leg

knuckle

hand and spring · spare ribs · belly

bacon, ham and gammon main meat cuts · prime collar · top back rashers · middle cut · short back rashers · oyster cut · long back rashers

middle collar

corner gammon

end collar

middle gammon

hock

fore slipper

fore hock butt · streaky · flank · gammon slipper

streaky rashers

small hock

joint[2] (*n*) meat butchered (p.98) for roasting (p.87) often still containing bone. Joints usually contain more than one muscle (p.95).

comminuted meat meat (p.95) removed from the low quality (p.128) parts of the carcass (p.98) after butchery (p.98), broken down by machines into very small pieces or viscous (p.23) suspensions (p.23). Comminuted meat is used in the manufacture of meat products (↓). **comminute** (*n*).

meat product fabricated food made from meat (p.95) or comminuted meat (↑), e.g. corned beef (↓), meat pies, sausages.

corned beef in the U.K., a canned (p.156), low quality beef comminute (↑). Outside the U.K., corned beef is cured (p.163) beef (↑).

lean (*adj*) of meat (p.95) having very little fat (p.52).

sheen (*n*) the shiny, bright surface of fresh meat (p.95)

taint (*n*) a bad odour (p.128) or taste (p.129) in food, particularly meat (p.95).

haematin (*n*) a group of metalloproteins (p.62) common in fish muscle (p.95), e.g. mackerel. Haematin proteins are rich in haem (p.79) compounds and cause rancidity (p.59) of oils (p.52) in stored fish due to oxidation (p.14).

fruit (*n*) the part of the plant which contains the
seeds (p.107). It is often edible (p.111) and rich
in vitamin C (p.74), also containing carbohydrate
(p.43) which occurs as glucose (p.46), fructose
(p.46), sucrose (p.46), starch (p.48), pectin
(p.49) and cellulose (p.49) among others.

vegetables (*n.pl.*) edible (p.111) plants, often rich
in vitamin C (p.74). Vegetables can be leaves,
e.g. cabbage, lettuce, spinach; roots, e.g.
parsnip, carrot; stems (↓) or stalks (↓), e.g.
celery, asparagus; flowers (↓), e.g. cauliflower,
calabrese or any other part except the fruit.
However, some fruits such as tomatoes and
cucumbers, and some seeds (p.107) such as
peas and beans are commonly called
vegetables.

**parts of plant
used as food**

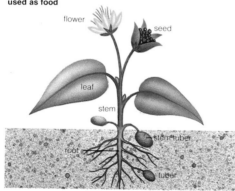

root (*n*) the part of the plant growing down into the
soil to support the plant and take up water and
nutrients (p.137). In some plants roots are able to
store food, e.g. carrot. Roots cannot bear leaves
and do not contain chlorophyll (p.30).

stem (*n*) the part of the plant, usually above the
ground which bears the buds (↓), leaves, flowers
(↓) and fruit, but it can be under ground, (*see*
rhizome (↓) and tuber (↓)). Some stems are green
and contain chlorophyll (p.30). *See also* shoot
(↓).

stalk (*n*) a stem (↑), e.g. celery, rhubarb.

corm (*n*) a thickened lower part of the stem (↑), capable of reproduction (p.107).

bulb (*n*) an organ (p.40) in some plants which consists of an underground axis with many thick overlapping leaves, for example, onions and garlic.

flower (*n*) the reproductive (p.107) part of the plant, often brightly coloured to attract insects. Some flowers are edible (p.111) and are commonly used as vegetables, e.g. cauliflower, calabrese.

tuber (*n*) an underground stem (↑) used for storing food, also able to produce new plants, e.g. potato, yam, cassava, sweet potato.

rhizome (*n*) a plant stem (↑) which grows under the ground producing shoots (↓) which grow into a whole new plant.

shoot (*n*) a young stem (↑).

bud (*n*) (1) a newly forming shoot (↑) or flower (↑) on a stem (↑). (2) a swelling which develops on a cell. This is part of a way of reproducing (p.107) called budding (p.151) found in micro-organisms (p.147) like yeast (p.151).

leaf (*n*) the part of the plant which gives off water and makes food by photosynthesis (p.43), e.g. lettuce, cabbage. Leaves contain a great deal of chlorophyll (p.30) and are usually thin, flat structures although some are specialized and have different shapes.

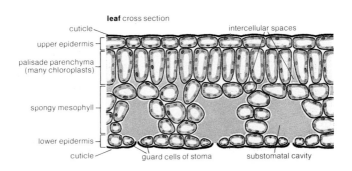

leaf cross section

cuticle

intercellular spaces

upper epidermis

palisade parenchyma (many chloroplasts)

spongy mesophyll

lower epidermis

cuticle

guard cells of stoma

substomatal cavity

parenchyma (*n*) the general name for plant tissues (p.32) having thin cell walls (p.32) usually with intracellular (p.29) spaces.

palisade parenchyma the layer of cells below the upper epidermis (↓) of a leaf.

palisade cell a cell in the palisade parenchyma (↑) of a leaf responsible for photosynthesis (p.43), which has many chloroplasts (p.30).

collenchyma (*n*) a plant tissue (p.32), with thick cellulose (p.49) cell walls (p.32), found in stems (p.104) and leaves. Collenchyma is used for support.

cortex (*n*) a many layered tissue (p.32) of the root or stem (p.104).

vascular bundle a tissue (p.32) in the leaf or stem (p.104) which enables the movement of substances from one part of the plant to another. It is made up of phloem (↓) and xylem (↓) cells (p.29).

xylem (*n*) the mostly dead tissue (p.32) in plants which forms long tubes connecting various parts of the plant and which enables water and nutrients (p.137) to be moved from the roots to the stem (p.104) and leaves.

phloem (*n*) the living tissue (p.32) which carries substances around the plant particularly the products of photosynthesis (p.43) from the leaves to other parts of the plant.

epidermis (*n*) the outer layer of cells on leaves, young roots and green stems (p.104).
  epidermal (*adj*).

parenchyma cells
in cross section

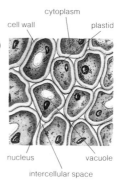

cytoplasm
cell wall
plastid
nucleus
vacuole
intercellular space

collenchyma cells
in cross section

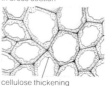

cellulose thickening

general structure of roots and stems

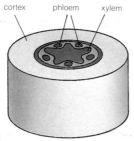

cortex   phloem   xylem

root

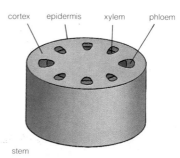

cortex   epidermis   xylem   phloem

stem

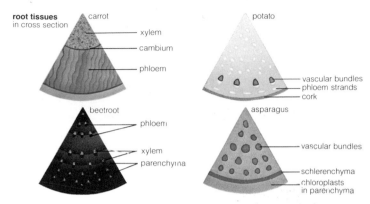

root tissues in cross section

carrot — xylem — cambium — phloem

potato — vascular bundles — phloem strands — cork

beetroot — phloem — xylem — parenchyma

asparagus — vascular bundles — schlerenchyma — chloroplasts in parenchyma

sclerenchyma (*n*) the hard, woody (p.108) tissue (p.32) found in stems (p.104) roots, leaves or fruit (p.104), used for support.

cork cambium the tissue (p.32) in the outer part of the stem (p.104) of a plant which makes the bark (↓).

phellogen (*n*) = cork cambium (↑).

bark (*n*) the outer layer of dead woody (p.108) tissue (p.32) in the stem (p.104) of a plant which protects the inner tissue.

embryo (*n*) a young plant contained inside the seed (↓). The word embryo is also used of young, unborn individuals in the animal as well as the plant world.

endosperm (*n*) the part of the seed (↓) which stores food for the young plant.

testa (*n*) the hard outer coat of a seed (↓) which protects the embryo (↑) and prevents water entering the seed until germination.

seed coat = testa (↑).

seed (*n*) the ripe product of reproduction (↓) in plants and the means by which the young can be spread. The seed contains the necessary material for a new plant to grow.

reproduction (*n*) the process by which all living organisms produce young like themselves. In plants, reproduction results in the formation of seeds (↑) from which new plants can grow.
reproduce (*v*), **reproductive** (*adj*).

scutellum (*n*) the part of the cereal (p.112) grain
(p.112) around the embryo (p.107) which is rich
in vitamins (p.72). The scutellum plus the embryo
is called the germ (↓).

germ (*n*) the part of the seed (p.107) made up of
the scutellum (↑) and the embryo (p.107).

lignin (*n*) an aromatic (p.11) compound found in
the cellulose (p.49) cell walls (p.32) of xylem
(p.106) and sclerenchyma (p.107) cells. Wood
consists mainly of lignin. **lignify** (*v*), **lignified**
(*adj*).

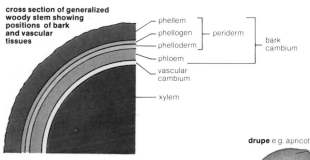

cross section of generalized
woody stem showing
positions of bark
and vascular
tissues

phellem
phellogen
phelloderm
— periderm
phloem
vascular cambium
xylem
bark cambium

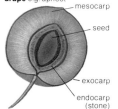

drupe e.g. apricot
mesocarp
seed
exocarp
endocarp (stone)

wood (*n*) hard tissue (p.32) containing lignin (↑),
made from dead xylem (p.106) cells which
supports the plant and carries water through it.
woody (*adj*).

pepo (*n*) a soft, juicy (p.90) fruit with hard outer
exocarp (↓) and many small seeds (p.107) at the
centre, e.g. melons, squashes.

drupe (*n*) a juicy (p.90) fruit in which the seeds
(p.107) are covered with a stony endocarp (↓)
called a pit (↓). Drupes usually have a fleshy
mesocarp (↓), e.g. apricot, peach.

berry (*n*) a juicy (p.90) fruit with many small seeds
(p.107), e.g. raspberry, mulberry.

citrus fruit a group of fruits having a thick exocarp
(↓), a high water content and a high citric acid
(p.21) content. Examples of citrus fruits include
lemon, orange, lime, grapefruit, tangerine,
citron.

hespiridium (*n*) = citrus fruit (↑).

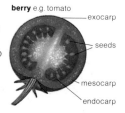

berry e.g. tomato
exocarp
seeds
mesocarp
endocarp

legume e.g. pea   **pome** e.g. apple

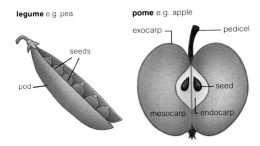

legume (*n*) a fruit of the family *leguminosae* (beans
    and peas) in which seeds grow inside a pod (↓).
pulse (*n*) the edible (p.111) seeds (p.107) of
    legumes (↑).
pod (*n*) a long, dry fruit which splits down the line
    where the margins of the carpel (↓) are joined,
    e.g. the pea pod.
pome (*n*) a fruit which grows from the receptacle
    (↓), not from the ovary (↓), such as an apple or
    pear.
carpel (*n*) the female reproductive (p.107) part of
    the flower containing the ovary (↓).
ovary (*n*) the inside of the carpel (↑) with a thick
    wall which grows into the fruit.
pedicel (*n*) the stalk of a single flower (p.105)
    where a shoot (p.105) bears flowers but no
    leaves.
pit (*n*) the stone from the cherry, peach, plum and
    apricot. The pit is made from the stony endocarp
    (↓) and surrounds the seeds (p.107). Oil (p.52)
    extracted (p.90) from pits is sometimes used in
    foods.
receptacle (*n*) the top part of the stalk (p.105) of
    any flower (p.105).
exocarp (*n*) the outer layer of tissue (p.32) on a fruit
    often called the skin.
epicarp (*n*) = exocarp (↑).
endocarp (*n*) the inner tissues (p.32) of a fruit
    which surround the seeds.
mesocarp (*n*) the fleshy and juicy layer of tissues
    (p.32) between the exocarp (↑) and endocarp (↑)
    in a fruit.

senescence (*n*) the process of growing older.
  senescent (*adj*).

harvest (*n*) the season for gathering crops.
  harvest (*v*).

ripe (*adj*) of fruit, ready for eating. **ripen** (*v*).

ethene (*n*) CH_2CH_2. A gas (p.22) produced by fruits
  which speeds up their ripening. Also known as
  ethylene.

cell death the death of some cells in an organism
  without necessarily resulting in the death of the
  whole organism.

necrosis (*n*) = cell death (↑).

climacteric (*n*) the time marking the beginning of
  the ripening of fruit when respiration (p.32)
  increases. Ethene (↑), which is produced by
  ripening fruits, controls the climacteric.
  climacteric (*adj*).

non-climacteric (*adj*) of fruit which do not show a
  climacteric (↑).

post-climacteric (*adj*) of climacteric (↑) fruit which
  are ripe and have completed the climacteric (↑).

CO_2 injury damage to fruit or vegetables caused
  by high carbon dioxide (CO_2) levels in storage.

CO_2 deprivation damage caused to fruit or
  vegetables by too low a level of carbon dioxide
  (CO_2) during storage.

H_2O vapour deficit injury or damage caused to
  fruit or vegetables during storage due to a low
  water vapour (p.24) content in the storage gas
  (p.22). Loss of turgor pressure (↓) results and the
  plant tissue (p.32) becomes limp and necrosis
  (↑) sets in.

transpiration (*n*) the movement of water up
  through the xylem (p.106) of the plant caused
  when water evaporates (p.24) from the stomata
  (↓) in the leaves.

sap (*n*) the liquid (p.22) transported in the xylem
  (p.106) and phloem (p.106) of a plant containing
  nutrients (p.137), minerals (p.16) and mainly
  water. Cell sap is the liquid in cell vacuoles
  (p.29).

stoma (*n*) a pore (p.32) on the under surface of a
  leaf. Stomata can open or close to control
  evaporation (p.24) of water from, and the entry of
  CO_2 into the leaf. **stomata** (*pl*).

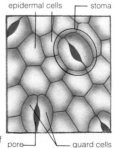

stomata
surface view of leaf

epidermal cells ⸻ stoma

pore⸻ ⸻ guard cells

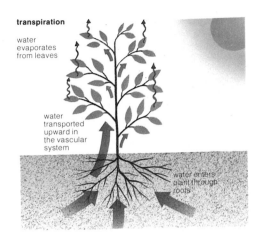

transpiration

water
evaporates
from leaves

water
transported
upward in
the vascular
system

water enters
plant through
roots

turgor pressure the internal pressure of a plant cell
which resists the inward elastic (p.115) pressure
of the cell wall (p.32). It is produced on the cell
wall by the water inside the cell. Turgor pressure
can be as much as nine times atmospheric
pressure.

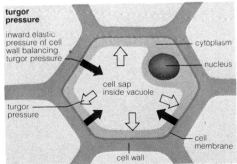

**turgor
pressure**

inward elastic
pressure of cell
wall balancing
turgor pressure

cytoplasm

nucleus

cell sap
inside vacuole

turgor
pressure

cell
membrane

cell wall

edible (*adj*) of any food suitable to be eaten.
inedible (*adj*) of any food or substance that is
unsuitable to be eaten.
papaya (paw-paw) (*n*) a large green-yellow fruit.

cereal (*n*) the edible (p.111) fruit of any grass which is used as food. The main cereals are wheat (↓), rice (↓), rye (↓), oats (↓), barley (p.114), maize (corn) (p.114), sorghum and millet.

grain (*n*) the name given to the edible (p.111) fruit of cereals (↑). The grain contains the embryo (p.107), scutellum (p.108), endosperm (p.107) and bran (p.119).

wheat grain

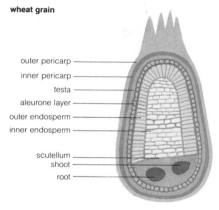

outer pericarp
inner pericarp
testa
aleurone layer
outer endosperm
inner endosperm
scutellum
shoot
root

wheat (*n*) one of the world's most important cereals (↑). There are many kinds of wheat but the three main kinds are *Triticum vulgare*, *Triticum durum* and *Triticum compactum*. These three are used for making bread, pasta (↓) and cakes respectively.

Durum wheat common name for *Triticum durum*.

pasta (*n*) food products made with Durum wheat (↑) including spaghetti, macaroni, vermicelli.

winter wheat wheat (↑) planted in the autumn and harvested (p.110) in the summer. These wheats generally have a high content of glutenin (p.115) and gliadin (p.115) and produce strong flour (p.119).

spring wheat wheat (↑) planted in the spring and harvested (p.110) in the late summer. These wheats generally have a lower content of glutenin (p.115) and gliadin (p.115) than winter wheat (↑) and produce a soft flour (p.116).

hard wheat can be either winter wheat (↑) or spring wheat (↑) and is used for bread making due to its high protein (p.61) content.

soft wheat can be either winter wheat (↑) or spring wheat (↑) and is used for cakes and softer goods due to its lower content of protein (p.61).

rice grain

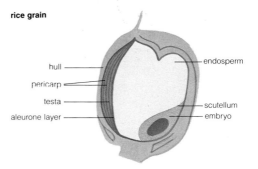

rice (*n*) the main cereal (↑) of the Eastern world. It may be grown on dry land (about 10% of the world harvest) or in standing water.

rye (*n*) a cereal (↑) grown mainly in Europe and Russia and used to make rye bread.

barley grain

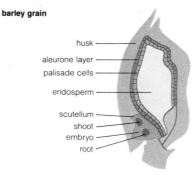

oats (*n*) a cereal (↑) mainly grown as animal feed but used in some countries in food products. It has no gluten (p.114) and so cannot be used for bread making.

oat grain

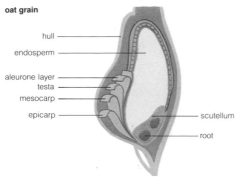

hull

endosperm

aleurone layer
testa
mesocarp
epicarp

scutellum

root

barley (*n*) a cereal (p.112) not normally milled
(p.116) due to its low protein (p.61) content.
Used in the production of beer (p.93).

maize (*n*) a cereal (p.112) grown mainly in America
and Africa. It is eaten as a whole grain (p.112) as
well as being milled (p.116) into cornflour (↓).

corn (*n*) = maize (↑).

maize grain
(corn)

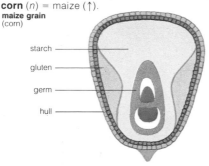

starch

gluten

germ

hull

cornflour (*n*) milled (p.116) corn (↑) containing
mostly starch (p.48) used in cooking (p.87) as an
extender (p.84), thickener (p.84) and to make
gels (p.25), e.g. in custard and blancmange.

gluten (*n*) a mixture of many proteins (p.61) of
wheat (p.112) flour (p.116) the main ones of
which are glutenin (↓) and gliadin (↓). It forms
during dough development (p.116) and is elastic
(↓) giving dough (↓) its springy properties.

glutenin (*n*) one of the two main proteins (p.61) of
wheat (p.112) flour (p.116). It contains many
disulphide bonds (p.65) and combines with
gliadin (↓) during dough development (p.116) to
make gluten (↑).

gliadin (*n*) with glutenin (↑) the main protein (p.61)
of wheat (p.112) flour (p.116). They combine to
form gluten (↑) during dough development (p.116).

gluten its formation during dough development

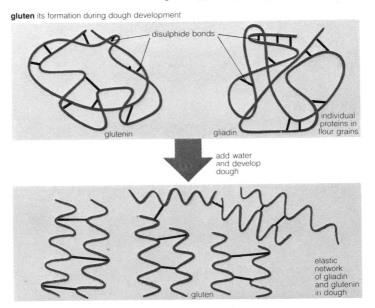

disulphide bonds

glutenin   gliadin

individual
proteins in
flour grains

add water
and develop
dough

elastic
network
of gliadin
and glutenin
in dough

gluten

elasticity (*n*) the way in which a solid (p.22) returns
to its original shape after it is stretched or
pressed. **elastic** (*adj*).

dough (*n*) the characteristically dry paste made
from cereal (p.112) flour (p.116) which is cooked
(p.87) to make breads and bread-like goods, e.g.
rolls. Yeast (p.151) may be added to dough in
order to act as a raising or leavening agent
(p.121). The yeast ferments (p.34) in the dough
producing carbon dioxide which causes the
dough to fill with small bubbles and so, to rise.

dough development simple yeast fermentation reaction

$$C_6H_{12}O_6 \longrightarrow 2C_2H_5OH \quad + \quad 2CO_2 \uparrow$$

glucose   ethanol   carbon dioxide

dough development the process in dough (p.115) making in which gluten (p.115) forms in the wheat (p.112) flour (↓). The yeast (p.151), which is added to the dough mixture, is allowed to ferment (p.34), produces carbon dioxide gas and the dough rises.

proving (n) the part of the process of dough development (↑) in which the mixed dough (p.115) is left in a warm place to allow the yeast (p.151) to ferment (p.34) producing bubbles of carbon dioxide (CO_2) gas and causing the dough to rise.

pastry (n) food or part of a food made mainly from flour (↓), water and fat (p.52) or oil (p.52). The lipid (p.51) component of pastry is usually called shortening (p.60) and the type and quality of this can greatly affect the type of pastry obtained.

flour (n) the powdered whole grain (p.112) or part of the grain of any cereal (p.112) produced by milling (↓).

milling (n) the process by which flour (↑) is made from cereal (p.112) grains (p.112). Also the process in which part of the grain is removed to leave a solid (p.22) product, e.g. removal of the bran (p.119) from rice to give white rice.

stone-milling a method of grain (p.112) milling (↑) in which the grain is powdered between two closely fitting stones. Only the top stone revolves and grinds (↓) the grain more finely as it moves to the outside of the stones.

grind (v) to break down to a powder.

roller-milling the most widely used method of making flour (↑) in which the grain (p.112) is ground between rolling drums.

fragmentation-milling a kind of roller-milling (↑) in which the white flour (p.120) is further divided into three groups of particle (p.24) size using a stream of air.

roller-milling

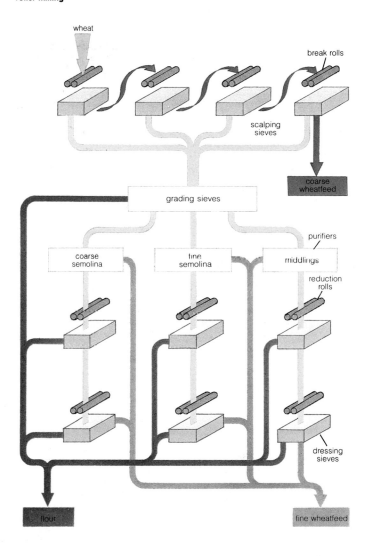

screening (n) the first process before milling (p.116) in which the grain (p.112) is separated from particles (p.24) larger and smaller than itself.

sorting (n) the second process before milling (p.116) in which the grain (p.112) is separated from other cereal (p.112) or grass grains.

scrubbing (n) the third process before milling (p.116) in which the dirt or dust is removed from the grain (p.112).

washing (n) the fourth process before milling (p.116) in which the grain (p.112) is washed and then the water is removed.

dockage (n) the material removed from grain (p.112) during the processes of screening (↑), sorting (↑), scrubbing (↑) and washing (↑).

cereal conditioning the fifth process before milling (p.116) in which the grain (p.112) is mixed with a small amount of water to make sure that the water content of the grain is correct. Either the wet grain is left for 24–72 hours or it is heated to 49°C for about 90 minutes. Conditioning allows easier removal of the bran (↓) during milling.

breaking (n) a process in milling (p.116) in which cleaned and conditioned (p.99) grain (p.112) is broken into small pieces by being passed through a set of break rolls (↓). By passing through a set of break rolls it is gradually powdered into flour (p.11) and the bran (↓) and germ (p.108) can be separated if required.

break roll one of the two grooved drums set very closely together through which grain (p.112) passes in the milling (p.116) process. Passage through the break rolls reduces (↓) the grain to pieces and then to coarse flour (p.116) called semolina (↓). Most mills have five sets.

semolina (n) the coarse flour (p.116) made of large particles (p.24) obtained from the break rolls (↑) in roller milling (p.116).

reduction[2] (n) the process by which the coarse particles (p.24) obtained from the breaking (↑) stage of milling (p.116) are further broken down to flour (p.116) by the reduction rolls (↓). The semolina (↑) is reduced to a fine powder (flour) and the germ (p.108) is removed. Large particles are sieved (↓) out and further reduced to flour.

preparation of wheat for milling

screening

dust and dirt
collected wheat grain

sorting

other seeds

wheat grain

scrubbing

washing

dust and dirt

conditioning

15% H₂O

to mill

reduction roll one of the two smooth drums set closely together through which the semolina (↑) from the break rolls (↑) passes. The reduction rolls reduce the semolina to fine flour (p.116). Each mill has a set of reduction rolls.

sieving (*n*) the process of removing large particles (p.24) from smaller ones, e.g. in milling (p.116), the large particles (p.24) are separated from the powdered endosperm (p.107). **sieve** (*v*).

bran (*n*) the outer layers of cereal (p.112) grain (p.112). Bran contains a high amount of dietary fibre (p.139) and some of the B vitamins e.g. thiamin (p.75), niacin (p.76).

mill stream the grain (p.112) pieces and flour (p.116) which pass out of the break rolls (↑).

pollards (*n*) the waste material remaining after milling (p.116) of cereal (p.112) grains (p.112). It is mainly made up of bran (↑) and small amounts of endosperm (p.107).

miller's offal = pollards (↑).

weak flour flour (p.116) containing only low amounts of protein (p.61) – about 8% by weight. Used mainly in production of cakes and pastry.

medium flour a flour (p.116) milled (p.116) from a mixture of wheats (p.112) with an average protein (p.61) content – about 10% by weight.

strong flour flour (p.116) containing high amounts of protein (p.61) – up to 17% by weight. Used in breadmaking where high protein, i.e. gluten (p.114), contents are needed to make a good dough (p.115) which rises well.

brown flour flours (p.116) such as wholemeal (↓), wheatmeal (↓), germ meal (↓), malted meal (p.120) which contain parts of the bran (↑) and germ (p.108) as well as powdered endosperm (p.107).

stone-ground flour flour (p.116) produced by the process of stone-milling (p.116). It usually contains the powdered whole grain (p.112), i.e. bran (p.112), germ (p.108) and endosperm (p.107).

wholemeal (*n*) brown flour (↑) containing 100% of the ground (p.116) grain (p.112).

wheatmeal (*n*) brown flour (↑) containing more than 85% of the powdered grain (p.112).

germ meal flour (p.116) containing added cooked (p.87) germ (p.108).

malted meal a flour (p.116) made from a mixture of wholemeal (p.119) and white flour (↓) with added malt flour (↓).

malt flour a flour (p.116) made from barley (p.114).

white flour a flour (p.116) which has had all of the bran (p.119) and germ (p.108) removed during milling (p.116). There are several kinds of white flour, e.g. patent flour (↓), straight run flour (↓).

extraction rate the percentage of the grain (p.112) obtained as flour (p.116), i.e. if 50 kilograms of flour is obtained from 100 kilograms of grain, the flour has a 50% extraction rate.

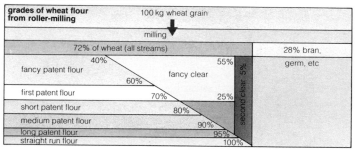

patent flour the finest kind of flour (p.116) obtained usually from the first three break rolls (p.118). It has an extraction rate (↑) of 25–40% and is used in bread and cake making.

straight run flour a general purpose flour (p.116) with an extraction rate (↑) of 70–75%, i.e. all of the powdered endosperm (p.107).

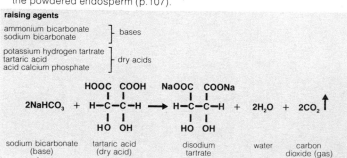

self-raising flour a medium flour (p.119) with added baking powder (↓) to help the aeration (p.57) of the dough (p.115).

baking powder a mixture of a dry acid (p.14) and a base (p.15) which, when water is added, react (p.12) to produce carbon dioxide (CO_2) aerating (p.57) the product.

raising agent = baking powder (↑).

leavening agent = baking powder (↑). Also used of yeast (p.151) which is added to flour (p.116) in breadmaking to produce carbon dioxide by fermentation (p.34).

high ratio flour flour (p.116) milled (p.116) from high quality wheat (p.112) with an extraction rate (↑) of less than 50%. Used in cake making as it can absorb (p.41) a large amount of water, fat (p.52) and sugar (p.44). It is usually heavily chlorinated (↓).

chlorination (*n*) the process by which a material is treated with chlorine (Cl_2).

aging (*n*) a chemical process used to oxidize (p.14) and bleach (↓) flour (p.116). Aged flour gives a more elastic (p.115) dough and better quality products.

maturation (*n*) = aging (↑).

improver (*n*) a substance or mixture of substances added to flour (p.116) to give a better product. Improvers are usually oxidizing agents and may oxidize (p.14) the flour (e.g. chlorine, sulphur dioxide), bleach (↓) the flour (e.g. benzoyl peroxide), or do both (e.g. chlorine dioxide).

bread structure

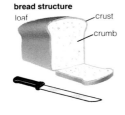

loaf — crust
— crumb

bleach (*v*) to whiten or remove colour. **bleaching** (*n*).

crust (*n*) the hard outer layer of bread produced during baking.

crumb (*n*) the solid structure within baked bread surrounding the trapped bubbles of carbon dioxide.

loaf volume the volume of a loaf of bread or of any bread product, e.g. rolls.

staling (*n*) the process in which the crumb (↑) of bread becomes hard and dry. Staling is due to the retrogradation (↓) of starch (p.48). **stale** (*adj*).

retrogradation (*n*) the process in which amylose (p.48) in the crumb (↑) of bread products slowly crystallizes (p.11) leading to staling (↑).

dairy products food products obtained from milk.
Milk is required in the U.K. to contain no less than
8.5% by weight of solids not including fat (p.52)
and not less than 3% by weight of fat. It is mainly
water (80–85%) but also contains protein (p.61),
carbohydrate (p.43), lipid (p.51), minerals (p.16)
and vitamins (p.72).

milk its percentage composition

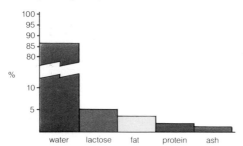

pasteurized milk milk in which most of the
vegetative (p.150) forms of bacteria (p.148),
mould (p.151) and yeasts (p.151) have been
killed by thermal processing (p.87). The milk is
usually heated at 63–65°C for 30 minutes or at
over 71°C for 15 seconds (the HTST (p.157)
method). Pasteurized milk should be stored at
4°C and has a shelf-life (p.155) of a few days.

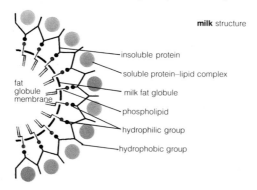

milk structure

insoluble protein

soluble protein–lipid complex

fat globule membrane

milk fat globule

phospholipid

hydrophilic group

hydrophobic group

homogenized milk milk forced through a narrow tube at 60°C to distribute the cream (↓) throughout the liquid (p.22).

sterilized milk milk which has been thermally processed (p.87) at high temperatures. Usually, it is treated at 104–110°C for 30–40 minutes. All micro-organisms are killed by this process and so the milk has a much longer shelf-life (p.155) than pasteurized milk (↑).

UHT milk ultra-high-temperature milk. Milk sterilized at 135–150°C for at least 1 second. Because of the short treatment time there are less colour and flavour changes in the milk than those brought about by the normal sterilization (p.155) process. The shelf-life (p.155) of UHT milk, when stored in special sealed cartons (p.94) can be very long (months).

uperization (*n*) a process by which milk is sterilized (p.155) by the injection of steam under pressure.

cultured milk products any milk product made by the addition of harmless micro-organisms (p.147), e.g. yoghurt (↓).

starter culture a mixture of bacteria (p.148) needed to carry out fermentation (p.34) in food products such as milk, for the production of yoghurt (↓) and cheese (p.124), and meat (p.95), for the production of fermented (p.34) meats, e.g. salami.

yoghurt (*n*) a fermented (p.34) milk product made by addition of a bacterial (p.148) culture (p.150) to any sort of milk. Most often used are *Lactobacillus bulgaricus* and *Streptococcus thermophilus,* which convert the lactose (p.47) in the milk into lactic acid (p.21) which gives the yoghurt its special flavour and helps to preserve (p.155) the product. The milk is homogenized (↑), heated at 88°C for about 30 minutes, cooled to 41–45.5°C, inoculated with the culture and held at this temperature until the yoghurt is ready at which time it is stored at 5–8°C.

cream (*n*) that part of milk in which the butter (p.60) fat (p.52) content is high. In natural, unhomogenized (↑) milk the cream floats at the top.

churn (*v*) to stir violently.

butyric acid $CH_3CH_2CH_2COOH$. A short chain fatty acid (p.53) found in triglycerides (p.53) of butter (p.60) and making up about 5–6% of their weight. Small amounts are also found in other fats.

butanedione (*n*) a chemical in butter (p.60) formed by bacteria (p.148) which gives the characteristic butter smell and flavour. Butanedione is added to margarine (p.60) to give it the flavour of butter. More commonly known as diacetyl.

diacetyl (*n*) = butanedione (↑).

cheese (*n*) a fresh or aged product made by coagulating (p.67) milk, cream (p.123), skim milk (↓), concentrated milk (↓), buttermilk (↓), dried milk or any combination of these with rennet (↓). The solids, called the curd (↓), are separated from the liquid (p.22) left after coagulation, called whey (↓), and are pressed to a firm cake to form the cheese. Most cheeses are fermented (p.34) with bacterial (p.148) cultures (p.150) over a period of time before being eaten. There are more than 400 known varieties of cheeses in the world.

whey (*n*) a watery liquid (p.22) separated from the curd (↓) in cheese making when the casein (↓) and most of the fat (p.52) have been removed. It contains about 1% protein (p.61), lactose (p.47), water soluble (p.72) vitamins (p.72) and minerals (p.16). Also known as lacto-serum.

curd (*n*) the coagulated (p.67) protein (p.61) formed when rennin (p.38), acid (p.14) or some other chemicals are added to fresh milk.

rennet (*n*) a commercial preparation of rennin (p.38), from the stomach (p.37) of calves (p.101), used in the production of some cheeses (↑). The rennet is added to the milk after it has been fermented with lactic acid (p.21) producing bacteria (p.148) to coagulate (p.67) the milk proteins (p.61).

skim milk milk from which the cream (p.123) has been removed.

filled milk a dehydrated (p.162) milk made from skim milk (↑) to which vegetable oil (p.52) has been added before the product is dried. Various levels of oil can be added depending on the intended use. Filled milk has a much better shelf-life (p.155) than dried whole milk.

concentrated milk milk which has been
concentrated (p.22) by evaporation (p.24) of
some of the water or by drying (p.162).
Some concentrated milks are sweetened.
evaporated milk = concentrated milk (↑).
Evaporated milks have no added sugar.
buttermilk (*n*) a by-product of the
manufacture of butter (p.60). It is made from
pasteurized (p.122) skim milk (↑) which has
been treated with an acid (p.14) butter
culture (p.150).
ice cream (*n*) a sweet mixture of milk, fat
(p.52), sugar (p.44), stabilizer (p.84),
emulsifier (p.84), flavouring and colouring
which is stored frozen (p.159) and eaten in
that form. Ice cream has a colloidal (p.25)
structure being a solid foam (p.26).
casein (*n*) the main protein (p.61) of milk. It
accounts for nearly 80% of all the protein in
milk. Casein is a mixture of several proteins
which combine together in milk to form
micelles (p.27) with calcium phosphate.

casein structure of micelle

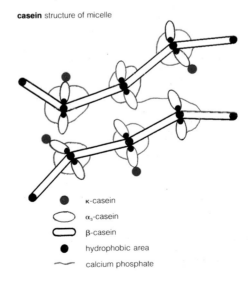

κ-casein

α_s-casein

β-casein

hydrophobic area

calcium phosphate

albumin (*n*) a simple protein (p.61) soluble (p.23) in water and coagulated (p.67) by heat, for example, lactalbumin (↓) in milk, ovalbumin (↓) in egg, and serum albumin in blood.

lactalbumin (*n*) a water soluble (p.23) albumin (↑) in milk; it makes up about 4% of the protein (p.61).

lactoglobulin (*n*) a water soluble (p.23) protein (p.61) in milk which makes up about 9% of the protein.

milk percentage content of different proteins

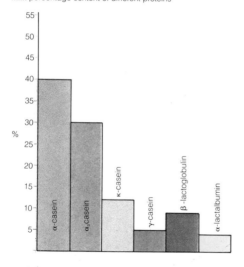

shell (*n*) the hard outer coating of eggs made of calcium carbonate ($CaCO_3$) and phosphate (p.20). It is permeable to gases (p.22) and stored eggs lose carbon dioxide (CO_2) through it. As bicarbonate is the main buffer in eggs this loss of CO_2 causes a pH rise (up to 9.0) in the egg white.

ovalbumin (*n*) the albumin (↑) of egg white which makes up 55% of the protein (p.61); it is a phosphoglycoprotein (p.52).

chalazae (*n.pl.*) the fibrous (p.65) protein (p.61) structures of the egg white which hold the yolk in place. **chalaza** (*sing.*).

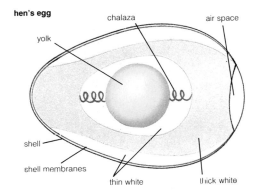

lysozyme (*n*) hydrolytic (p.13) enzyme (p.68) in egg white. Can cause thinning of the egg white by breaking down proteins (p.61).

ovomucoid (*n*) a glycoprotein (p.62) of the egg white which makes up about 11% of the protein (p.61).

ovomucin (*n*) an egg glycoprotein (p.62) which is thought to make the white thick. Ovomucin and lysozyme (↑) may react (p.12) together to cause the thinning of egg white during storage.

avidin (*n*) a protein (p.61) found in egg white. Avidin is able to bind the vitamin (p.72) biotin (p.77) very strongly making it biologically unavailable (*see* biological availability (p.138)).

lipovitellin (*n*) one of the two main lipoproteins (p.62) of egg yolk.

lipovitellenin (*n*) the second main lipoprotein (p.62) in egg yolk.

processed egg eggs which have been treated and stored in one of several different ways. Eggs can be frozen (p.159) or dried (p.162).

chocolate (*n*) a substance made from husked fermented (p.34) and roasted (p.87) cocoa beans (p.94) which are then refined (p.90) and mixed with sugar (p.44), cocoa butter, flavouring, lecithin (p.55) and, for milk chocolate, milk solids.

UHTS (ultra-high-temperature-sterilization). A method of sterilizing (p.155) milk at temperatures above 135°C. *See* UHT (p.156).

main sensory organs of taste and smell

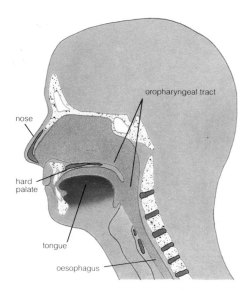

quality (n) a characteristic or property of a substance or food. Quality in foods is made up of all the characteristics of the parts of that food, e.g. taste, odour (↓), appearance, texture, etc.

qualitative (adj) of a characteristic which depends on quality.

quantitative (adj) of a characteristic which depends on the amount (quantity) of a substance or thing.

sense (n) any of the ways in which an organism interacts (p.16) with the outside world. In humans these are touch, smell, taste, sight, hearing.

tactile (adj) of the sense of touch.

odour (n) a smell.

aroma (n) = odour.

mouthfeel (n) the feeling of a food when in the mouth. Mouthfeel is usually used to indicate texture (p.27) but it can also be affected by stimuli (p.130) such as tingling, hotness, irritation and coolness.

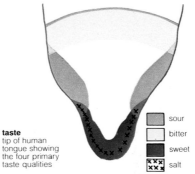

taste
tip of human
tongue showing
the four primary
taste qualities

sour
bitter
sweet
salt

taste (*n*) the sense (↑) by which the flavour of a
food or substance is distinguished or observed in
the mouth. Taste is made possible by the taste
buds (↓) on the tongue. Taste is also used to
describe the actual combination of flavours of a
food. In food quality (↑) analysis there are said to
be only four basic tastes, sour, sweet, bitter and
salt all sensed by taste buds on different parts of
the tongue. (*v*)

bitter (*n*) *see* taste (↑).

salt² (*n*) *see* taste (↑). **salty** (*adj*).

sweet (*n*) *see* taste (↑).

sour *see* taste (↑).

taste bud (*n*) a special kind of cell found in the
tongue with receptors (p.130) on its surface. The
four basic tastes in foods (i.e. sweet, sour, bitter
and salt), can be recognized by taste buds.
Taste buds are connected to the brain by nerves.

taste bud

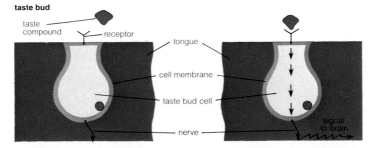

taste
compound
receptor
tongue
cell membrane
taste bud cell
nerve
signal
to brain

receptor (*n*) a protein (p.61) or other macromolecule (p.10) on the surface of a cell which can interact (p.16) with and recognize certain substances or molecules. Receptors on the surface of taste bud (p.129) cells interact with certain food flavour compounds and send signals through the cell and the nerves to the brain. Receptors are also present on the surface of cells in the oropharyngeal tract (throat) and are involved in sensing aroma (p.128).

organoleptic (*adj*) of those properties of a food concerned with stimulation (↓) of the senses (p.128).

stimulus (*n*) something which brings about activity. Stimuli are events or substances which affect the senses (p.128). **stimuli** (*pl.*), **stimulate** (*v*), **stimulation** (*n*).

eye appeal the appearance of a food. Those qualities in a food (e.g. colour, shape), which stimulate (↑) the sense (p.128) of sight.

subjective (*adj*) a personal opinion not based on fact. **subjectivity** (*n*).

objective (*adj*) an opinion based on facts. **objectivity** (*n*).

quality control the monitoring and control of food quality from the farm to the consumer (↓) to ensure a uniform high quality.

grading peas by size

peas from washer

oversize peas and rubbish

large peas
canned processed peas

medium peas
canned garden peas

small peas drying

grade (*v*) to arrange in groups. Food is often graded by size, colour and general quality. **grade** (*n*), **grading** (*n*).

characterization (*n*) a method by which properties and characteristics are discovered. This may be done by chemical analysis (p.20), physical analysis (p.133) or, in the case of foods, by sensory analysis (↓).

sensory analysis a method by which the organoleptic (↑) properties of a food or food product are determined. Sensory analysis uses human tasters as the means by which the properties are measured. The human tasters may be consumers (↓) or a taste panel (↓).

consumer (*n*) a buyer or user of a product.

taste panel a group of people who test a food with their own senses (p.128). Taste panellists may be either trained (↓) to test the food or not trained.

trained taste panel a small group of people who have been trained to taste and, thereby test, a food or range of foods. Trained taste panels usually consist of six to ten people.

discrimination test sensory analysis (↑) in which the difference between food samples are judged, e.g. paired comparison test (↓), triangle test (↓), duo-trio test (↓), taste threshold test (↓).

difference test = discrimination test (↑).

paired comparison test sensory analysis (↑) in which a person is asked if there is a difference between two samples.

paired stimuli test = paired comparison test (↑).

triangle test sensory analysis (↑) in which two samples are the same and one is different and the judge is asked to find the odd one out.

duo-trio test a test similar to the triangle test (↑) in that three samples are used. In this case the person is asked to say which of two samples is different from a control sample (↓).

control sample a sample with known characteristics.

taste threshold test sensory analysis (↑) to determine the lowest concentration (p.22) of a substance which can be tasted.

descriptive test sensory analysis (↑) in which the characteristics of a food are described by the judge or taste panel (↑).

descriptor (*n*) a word used to describe a quality in a food sample during a descriptive test (↑). Descriptors are usually agreed by the judges before a test.

single sample test sensory analysis (↑) in which the characteristics of a sample are tested by one person who may be a consumer (↑) or an expert.

descriptors
in descriptive tests
of chicken legs

wrinkled skin
bright skin
chewy
rancid odour
rancid flavour
fresh chicken taste
bland taste

ranking test sensory analysis (p.131) in which judges are asked to show their order of preference (↓) of a certain character in a sample.

preference (*n*) a liking of one thing better than another.

scoring test sensory analysis (p.131) in which judges are asked to show the differences between a set of samples by giving a score (↓) to each sample.

score (*v*) to give a numerical value or point to an item so that it may be compared with other items. **score** (*n*).

flavour profile sensory analysis (p.131) in which the flavour of a food sample is defined.

texture profile sensory analysis (p.131) in which the texture of a food sample is judged.

QDA quantitative descriptive analysis. Sensory analysis (p.131) in which several characteristics of a food are judged and either scored (↑) or ranked (↑).

acceptance-preference test sensory analysis (p.131) in which the acceptibility or desirability of the food product is determined. These kinds of test are commonly used with consumers (p.131).

hedonic scale simple pictorial scale

dislike   like

hedonic scale a scale of words or phrases or even pictures used to show preference (↑) for a food sample in an acceptance-preference test (↑). Phrases like 'dislike extremely' and 'like extremely' are used by the judge to describe the sample.

standardization (*n*) a process in which samples or objects are judged against an accepted example or standard. Also any method used to change a sample or result in order to make it standard. **standard** (*n*), **standardize** (*v*).

optical microscope

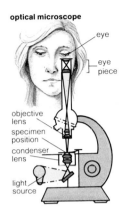

criterion (*n*) a standard used to make judgements.
criteria (*pl*).

physical analysis a method, using instruments of
measurement, of determining the character of
food, e.g. texture (p.27), elasticity (p.115).

microscopy (*n*) a way of looking at micro-
organisms (p.147) and the small structures of
foods by magnification (↓) using instruments
such as light microscopes (↓) and electron
microscopes (↓).

magnify (*v*) to make the size of an object or
structure appear larger. **magnification** (*n*).

light microscope an instrument which magnifies
(↑) objects up to hundreds of times their actual
size using a light beam.

electron microscope an instrument which can
magnify (↑) objects up to over a million times
their actual size using a beam of electrons (p.9).

Warner-Bratzler shear apparatus an instrument
for measuring the tenderness (p.99) of meat
(p.95). The machine measures the force (p.134)
needed to tear a piece of meat against a
triangular opening.

penetrometer (*n*) an instrument used to measure
the textural (p.27) quality of meat (p.95) or plant
foods. The machine measures the depth to which
a pointed device can enter a food sample when a
particular force (p.134) is applied.

electron microscope

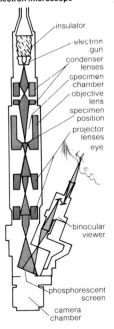

penetrometer

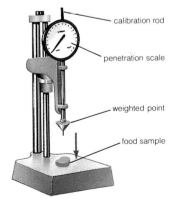

tenderometer (*n*) an instrument like the penetrometer (p.133) used to measure the tenderness (p.99) of a food.

texturometer (*n*) an instrument used to measure the texture (p.27) of a food. A tenderometer (↑) is an example of a texturometer.

Magness Taylor pressure tester an instrument for measuring vegetable texture (p.27). The machine measures the force (↓) needed to puncture a food sample to a particular depth.

farinograph (*n*) an instrument used to measure the stability and consistency of dough (p.115). The machine measures the force (↓) needed to turn a set of blades at a set speed in a dough mixture.

force (*n*) power, strength; that which leads to a change in a physical system.

shortometer (*n*) an instrument used to measure the breaking strength of pastries (p.116) and other baked foods. Not now used very often.

viscometer (*n*) an instument used to measure the viscosity (p.27) of liquids (p.22) and pastes.

absorptiometer (*n*) an instrument used to measure the amount of light passing through a solution (p.22). As coloured substances absorb (p.41) light the instrument can be used to measure the amounts of such substances in solutions.

spectrophotometer (*n*) an instrument which measures the amount of ultra-violet or visible light absorbed (p.41) by a substance. Most coloured compounds absorb light of specific wavelengths and so can be characterized.

refractometer (*n*) an instrument used to check the concentration (p.22) and purity of solutions (p.22) of sugars (p.44), oils (p.52) and other food compounds.

pH meter an electronic instrument which measures the pH (p.14) of a solution (p.22).

chromatography (*n*) methods used to separate different kinds of molecules and macromolecules (p.10) from one another. Most forms of chromatography rely on being able to attract solutes (p.22), which are the molecules to be separated, from a solvent (p.22) using a specially prepared solid material, e.g. paper in paper chromatography (↓) or silica in TLC (↓) or HPLC (↓).

gel filtration chromatography chromatography (↑) in which molecules or macromolecules (p.10) are separated according to their size.

gel permeation chromatography = gel filtration chromatography (↑).

size exclusion chromatography = gel filtration chromatography (↑).

ion exchange chromatography chromatography (↑) in which molecules and macromolecules (p.10) are separated depending on their charge.

thin layer chromatography the separation of molecules on thin layers of particulate materials like silica. The solvents (p.22) to be used depend on the molecules to be separated. **TLC** (*abbr*).

paper chromatography
separation of pigments

- strip of absorptive paper
- carotene
- pheophytin
- xanthophyll
- chlorophyll a   unknown substance 1
- chlorophyll b   unknown substance 2
- concentrated chlorophyll extract

solvent

paper chromatography the separation of molecules on sheets of specially prepared paper. As in TLC (↑), solvents (p.22) are chosen to give a good separation of the molecules of interest.

high-performance liquid chromatography a form of chromatography (↑) which uses rigid, incompressible particulate materials (e.g. silica) in sealed steel cylinders so that high pressures may be used to force solvents (p.22) and solutes (p.22) through. Gel filtration chromatography (↑) and ion exchange chromatography (↑) may be carried out by HPLC at much higher speeds and often with much better results. **HPLC** (*abbr*).

digestibility (*n*) the amount of food which is absorbed (p.41) from the digestive tract (p.42) into the blood stream, usually about 90–95%. Digestibility is measured as the difference between intake of food and output of waste from the body.

nitrogen balance

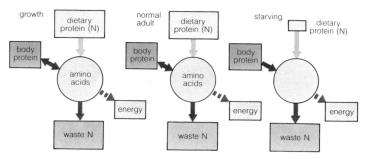

nitrogen balance the difference between the intake of nitrogen in the diet (↓), and the loss of nitrogen in waste products. Healthy adults lose the same amount as they take in, this is called *N equilibrium*. During growth the intake is greater than the loss, this is called *positive N balance*. In some diseases the loss is greater than the intake and is called *negative N balance*.

N equilibrium *see* nitrogen balance (↑).

positive N balance *see* nitrogen balance (↑).

negative N balance *see* nitrogen balance (↑).

RDA the recommended daily amount or allowance of nutrients (↓) taken daily.

reference protein the perfect protein (p.61), which does not actually exist but is a useful standard with which to compare real proteins. 100% of a perfect protein would be used in the body because it would contain the perfect balance of amino acids (p.61) needed for growth and repair. The proteins most like a reference protein are egg and human milk proteins where 90–95% of the protein is used by the body when fed at low levels in the diet (↓). It allows a measure of the usefulness of real proteins in the diet.

adequate diet

carbohydrate

vitamins

minerals

protein, fat

dietetic foods

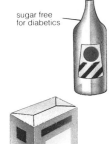

sugar free
for diabetics

gluten free foods
for people with
gluten allergy

diet (*n*) the food eaten.

adequate diet a diet (↑) which provides all the
necessary nutrients (↓) and energy in the correct
quantities to give good health and an acceptable
body weight.

balanced diet a diet (↑) in which each nutrient (↓)
is supplied in the correct quantities relative to all
other nutrients.

dietitian (*n*) a person who plans meals and special
diets (↑) for individuals or groups and teaches
about nutrition (↓).

dietetic food food prepared for people with special
nutritional (↓) needs such as people suffering
from illness or healthy people who may have
special needs.

nutrients (*n. pl.*) vitamins (p.72), minerals (p.16),
amino acids (p.61) and fats (p.52) which are
necessary in the diet (↑). As sources of energy
are not called nutrients an expression often used
is 'energy and nutrients'.

nutrition (*n*) the study of foods in relation to the
needs of the living creature. Nutrition is also used
to describe the diet (↑) of a particular person.
nutritional (*adj*).

nutrification (*n*) the addition of nutrients (↑) to
foods in large quantities.

nutritionist (*n*) a person whose job it is to use the
science of nutrition (↑) to improve health and
control food-related disease.

biological availability

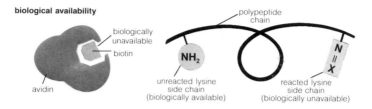

biologically
unavailable
biotin

avidin

polypeptide
chain

NH₂

unreacted lysine
side chain
(biologically available)

reacted lysine
side chain
(biologically unavailable)

biological availability the amount of a nutrient
(p.137) in a food is obtainable for use by the
body. For example, the egg white protein (p.61)
avidin (p.127) is able to bind the vitamin (p.72)
biotin (p.77) making it completely biologically
unavailable; the essential amino acid (p.63)
lysine can undergo Maillard reaction (p.92)
making such reacted lysines in a protein
biologically unavailable.

nutritive ratio the value of a food for growth
compared to its fattening value. It is the sum of
the digestible (p.41) carbohydrate (p.43),
protein (p.61) and 2.3 × fat (p.52) divided by
digestible protein.

index of nutritional quality the overall nutritional
(p.137) value of a food. **INQ** (*abbr*).

malnutrition (*n*) illness caused by poor or
inadequate nutrition (p.137).

fast (*v*) to go without food. **fast** (*n*), **fasting** (*n*).

ACH index arm, chest, hip index; where the arm
circumference, chest diameter and hip width is
measured to judge the state of nutrition (p.137)
of a person.

blood sugar the level of glucose (p.46) in the
blood which provides the energy for muscular
(p.95), metabolic and brain activity.

crude fibre the indigestible (p.41) part of food left
after other nutrients have been digested (p.41)
and absorbed (p.41) in the body. Crude fibre in
foods can be measured after extraction (p.90)
with dilute (p.23) acid (p.14) followed by dilute
alkali (p.15). In recent years fibre has been shown
to be necessary in the diet (p.137). Diets low in
fibre are thought to lead to such illnesses as
heart disease (p.140) and atherosclerosis (p.140)
as well as diseases of the alimentary canal (p.42).

fibre (*n*) = crude fibre (↑).
roughage (*n*) = crude fibre (↑).
reference man an imaginary man used for
    determining nutrient (p.137) and energy needs.
    He is 5 feet 9 inches tall (1.72 metres), weighs
    154 pounds (70 kilograms) and is aged between
    twenty-three and fifty.
reference woman an imaginary woman used for
    determining nutrient (p.137) and energy needs.
    She is 5 feet 5 inches tall (1.62 metres), weighs
    128 pounds (58 kilograms) and is aged between
    twenty-three and fifty.

fat fold test

fatfold test a test of body fatness by measuring the
    thickness of a fold of skin on the arm, back or
    other part of the body.
skinfold test — fatfold test (↑).
ectomorph (*n*) a tall, thin person who may have
    muscles (p.95) which are not fully formed. Not a
    very widely used word in modern nutrition (p.137).
endomorph (*n*) a short, well-built person. Not a
    very widely used word in modern nutrition (p.137).
mesomorph (*n*) a person who is well-built with fully
    formed muscles (p.95). Not a very widely used
    word in modern nutrition (p.137).
Benedict-Roth spirometer equipment used to
    measure the amount of oxygen used by a person
    (or the amount of carbon dioxide (CO_2)
    produced) from which the amount of energy used
    can be determined.

heart disease the main cause of early death in the Western world due, in part, to a diet (p.137) high in lipids (p.51) and refined (p.90) foods and low in fibre (p.139).

hypertension (*n*) a disorder in which a person has abnormally high blood pressure. Hypertension has been linked to dietary (p.137) factors especially a high intake of salt (NaCl).

development of atherosclerosis

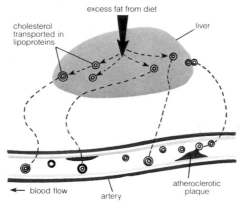

excess fat from diet

cholesterol transported in lipoproteins

liver

← blood flow

artery

atheroclerotic plaque

atherosclerosis (*n*) a disease of the arteries (blood vessels) in which a build up of lipids (p.51), particularly cholesterol (p.54), and connective tissue (p.97) leads to blockages. Eventually, the disease leads to death.

overweight (*n*) a condition in which the body weight is up to 10% higher than normal.

obesity (*n*) a condition in which the body weight is far greater than normal. Weights of 15–25% over the normal range for a person are considered obese.

adipose tissue tissues (p.32) containing cells filled with lipids (p.51) stored in vacuoles (p.29). It is the growth of adipose tissue which causes obesity (↑). However, adipose tissue is a normal part of every animal body and is necessary for storing lipids as a source of energy.

metabolic disorder a disorder or disease caused by the lack of an enzyme (p.68) or similar factor in a person which leads to the impairment of their normal metabolism (p.144), e.g. the lack of an enzyme normally used in the breakdown of the amino acid (p.61) phenylalanine in phenyl-ketonuria (↓). Metabolic disorders are inherited from the parents of affected people and may lead to serious symptoms (p.143). The damaging symptoms of many metabolic disorders can be overcome by careful control of dietary (p.137) intake of foods containing compounds which are involved in that part of the person's metabolism which has been upset by the disorder.

phenylketonuria (n) a metabolic disorder (↑) in which the enzyme (p.68) phenylalanine hydroxylase is lacking or inactive and therefore the amino acid (p.61) phenylalanine is not metabolised (p.144).

anorexia nervosa a psychological (↓) illness resulting in great loss of weight due to a refusal to eat.

psychological (adj) of the mind.

kwashiorkor (n) a disease of young children whose diet (p.137) lacks protein (p.61). It results in a lack of growth. Kwashiorkor is not normally seen since marasmus (↓) or true starvation is more common.

marasmus (n) starvation. This differs from kwashiorkor (↑) in that all nutrients are missing from the diet (p.137). If food is not given marasmus results in death.

marasmus

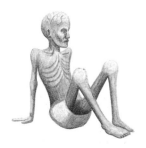

protein-calorie-malnutrition the name given to the range of diseases resulting from the lack of various different nutrients (p.137) from the diet (p.137). PCM is the most common malnutrition (p.138) problem in the world. **PCM** (*abbr*).

alcoholism (*n*) an illness in which alcohol (p.93) is drunk in large quantities resulting in dependence on it, damage to the liver (p.41) and death.

ketosis (*n*) a build up of ketones (−C=O) in the blood. Ketones are produced during the metabolism (p.144) of lipids (p.51) when there is a lack of carbohydrate (p.43) in the diet (p.137).

ketosis
fasting

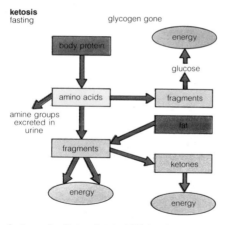

ketogenic diet a diet (p.137) low in carbohydrate (p.43) which causes ketosis (↑).

cancer (*n*) the second most common cause of early death in the Western world. Cancer is a group of diseases in which cells reproduce (p.107) themselves out of control. It has been linked with dietary (p.137) patterns in recent years.

carcinogen (*n*) a substance known to cause cancer (↑).

mutagen (*n*) a substance known to cause damage to the DNA (p.31). Through reproduction (p.107), this causes damage to the embryo (p.107).

teratogen (*n*) a substance known to cause damage to the embryo (p.107) before birth.

human embryo

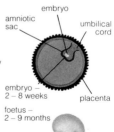

embryo at eight weeks

food allergy an unusual bodily response in a person to a food or foods that do not produce a response in most people. Many effects or symptoms (↓) have been described in people sensitive to food allergens (↓) including disorders of the alimentary canal (p.42), skin and lungs. Other symptoms, including hyperactivity (↓) in children, have been reported but are not so well understood.

allergen (*n*) a non-toxic (p.41) substance or particle (p.24) which causes slight or severe illness (food allergy (↑)), in people sensitive to it. Many food compounds have been suggested to be allergens.

hyperactivity (*n*) a state of being overactive. A condition, especially in children, in which the person cannot keep still and needs to be active or constantly moving. It has been associated with food allergies (↑). **hyperactive** (*adj*).

xenobiotic (*n*) a substance foreign to the body, e.g. many synthetic (p.13) food additives (p.84).

unhealthy foods

fatty food: fried egg, chips, lard, salt, salami, fried bread, fatty meat, fatty sausages, butter, white bread

sugary food: white sugar, sweets, very refined high sugar content convenience foods high in additives, ice cream, double cream, sugary drink

health food foods said to be natural and good for the body. Health foods contain no additives (p.84) and are not refined (p.90).

vegetarian (*n*) a person who does not eat meat (p.95)

lacto-ovo-vegetarian (*n*) a vegetarian (↑) who eats animal products, e.g. milk, eggs.

vegan (*n*) a person who does not eat meat (p.95) or any animal product.

symptom (*n*) a change in the body which shows its state of health or illness.

metabolism (*n*) all of the reactions (p.12) which take place in an organism or cell. The two main parts of metabolism are catabolism (↓) and anabolism (↓). **metabolize** (*v*), **metabolic** (*adj*).

metabolite (*n*) any compound produced during metabolism (↑).

metabolic pathway of a chain of reactions (p.12) each following the other and using the product of the first as the substrate (p.70) of the second.

catabolism (*n*) the breakdown of food molecules inside the cell to produce energy and the compounds used in synthesis (p.13).

anabolism (*n*) the building of new molecules and macromolecules (p.10) inside the cell using the energy and building block compounds produced by catabolism (↑).

catabolism and anabolism

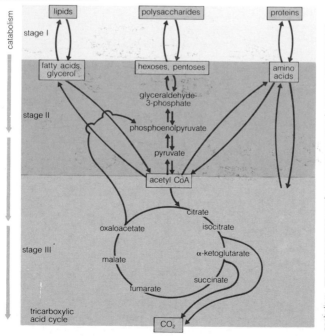

metabolism chemical reations in a plant cell

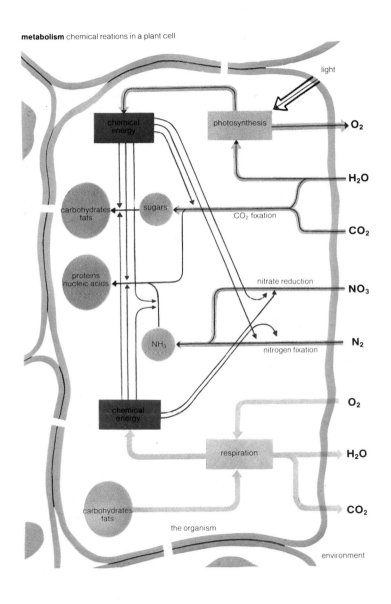

feedback (*n*) a process by which a product of a metabolic pathway (p.144) either speeds up or slows down reactions (p.12) earlier in the pathway.

metabolic rate the rate at which enegy is used up in the body.

basal metabolic rate the rate of energy output of a body after a twelve hour fast (p.138). The basal metabolic rate is also defined as the amount of energy needed by the body at rest in the fasting state. BMR is a good indication of the energy needed to sustain the life process in a person. **BMR** (*abbr*).

resting metabolic rate the energy used by the body at rest after eating or exercise. RMR is normally measured in preference to BMR as it is easier to measure. **RMR** (*abbr*).

basal metabolism the continuous metabolic (p.144) work done by the cells of an organism.

SDA specific dynamic activity. The energy used to assimilate (p.41) nutrients (p.137) from food.

SDE specific dynamic effect = SDA (↑).

net dietary protein energy ratio the protein (p.61) content of a diet (p.137) or food expressed as the ratio of protein energy to total energy multiplied by net protein utilization (p.67).

protein efficiency ratio a measure of protein (p.61) quality which is calculated from the extent to which any one protein leads to weight gain in a growing child. **PER** (*abbr*).

joule (*n*) the unit of energy which has replaced the calorie (↓) in the International System of Units. 4.184 joules (J) = 1 calorie.

calorie (*n*) a unit of measurement of the energy released when a nutrient (p.137) or food is oxidized (p.14). It has now been replaced by the joule (↑).

empty calories of foods that only supply energy with very few nutrients (p.137).

energy (*n*) work or the ability to do work. Energy in the body is obtained from chemicals and stored as ATP (p.36) and is used for the work of movement, to produce heat and in other chemical reactions (p.12) such as synthesis (p.13).

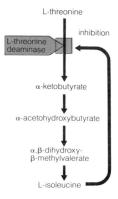

feedback inhibition of an enzyme by the final product in a metabolic chain

L-threonine

L-threonine deaminase

inhibition

α-ketobutyrate

α-acetohydroxybutyrate

α,β-dihydroxy-β-methylvalerate

L-isoleucine

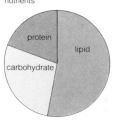

energy provided by equal weights of the three major nutrients

protein

lipid

carbohydrate

colonial green algae

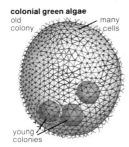

old colony
many cells
young colonies

bacteriophage attacking a bacterium

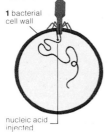

1 bacterial cell wall

nucleic acid injected

2 parts of new bacteriophages synthesized in bacterial cell

3 bacterium destroyed, new bacteriophages released

microbiology (*n*) the study of micro-organisms (↓).

micro-organism (*n*) organisms which are very small, usually containing only one cell and which cannot be seen with the human eye. There are five kinds of micro-organism; (1) protozoa (↓), (2) algae (↓), (3) viruses (↓), (4) bacteria (p.148), and (5) fungi (p.150), e.g. moulds (p.151).

microbe (*n*) = micro-organism (↑). **microbial** (*adj*).

protozoa (*n.pl.*) single-celled (↓) animals most of which are harmless to man. Some protozoa cause food poisoning (p.152), e.g. *Toxoplasma gondii* found in raw meat (p.95).

single-celled (*adj*) of organisms containing only one cell.

unicellular (*adj*) = single-celled (↑).

multicellular (*adj*) of organisms made up of many cells.

viruses

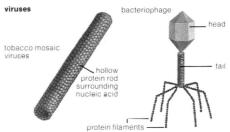

bacteriophage
head
tail
tobacco mosaic viruses
hollow protein rod surrounding nucleic acid
protein filaments

algae (*n. pl.*) simple plants which can be either single-celled (↑) or multicellular (↑). *Chlorella* is a single-celled alga which is used to make single cell protein (p.67). Seaweeds (↓) are multicellular algae. **alga** (*sing.*).

seaweed (*n*) a general name for any large algae (↑) in the sea.

virus (*n*) the smallest of all organisms most of which can only be seen in the electron microscope (p.133). They have no cell structure and only contain nucleic acid (p.31) surrounded by a coat of protein (p.61). Viruses can only reproduce (p.107) inside another living cell and are therefore always pathogenic (p.152).

bacteriophage (*n*) a virus (↑) which reproduces (p.107) inside bacterial (p.148) cells.

bacteria (*n. pl.*) a large group of micro-organisms (p.147) which can be both harmful and helpful to man. They are single-celled (p.147) but contain no nucleus (p.30). Many are pathogenic (p.152) and cause food poisoning (p.152) as well as food spoilage (p.154) but some live in the intestines of man and provide benefits such as producing vitamins (p.72), e.g. biotin (p.77), and others are useful in fermenting (p.34), e.g. yoghurt (p.123). Bacteria are divided into four main classes; (1) coccus (↓), (2) bacillus (↓), (3) vibrio (↓), (4) spirillum. **bacterium** (*sing.*), **bacterial** (*adj*).

coccus (*n*) a group of bacteria (↑) with a spherical, ball-like shape. **cocci** (*pl*).

bacillus (*n*) a group of bacteria (↑) with a rod-like shape. **bacilli** (*pl*).

vibrio (*n*) a group of bacteria (↑) with a short, curved comma-like shape.

spirillum (*n*) a group of bacteria (↑) with a long, coiled shape.

Gram stain a coloured compound used to stain (↓) some bacteria (↑) for microscopy (p.133). It is used to distinguish two groups of bacteria, those coloured by the stain (Gram positive (↓)) and those not coloured (Gram negative (↓)).

stain (*adj*) to react (p.12) coloured compounds with other materials so that those materials become coloured themselves.

gram positive a class of bacteria (↑) which are coloured violet by the Gram stain (↑).

gram negative a class of bacteria (↑) which are not coloured violet by the Gram stain (↑).

halophilic bacteria bacteria (↑) able to grow in high concentrations (p.22) of salts (p.16).

halophile (*n*) a halophilic bacterium (↑).

thermophilic bacteria bacteria (↑) able to grow at high temperatures (55–65°C).

thermophile (*n*) a thermophilic bacterium (↑).

thermoduric (*adj*) of bacteria (↑) which can stand the effects of high temperatures.

psychrophilic bacteria bacteria (↑) which grow at 20–30°C optimum and even below 0°C.

psychrophile (*n*) a psychrophilic bacterium (↑).

psychroduric (*adj*) of bacteria (↑) which can stand the effects of low temperatures.

bacteria

bacilli

spirochaetes

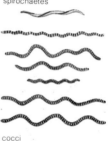

cocci

diagram showing ranges of temperature for growth and resistance of some bacteria

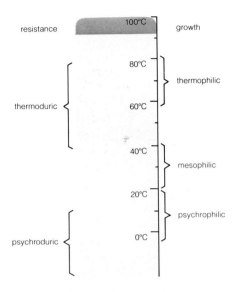

resistance — 100°C — growth

80°C

thermophilic

thermoduric

60°C

40°C

mesophilic

20°C

psychrophilic

0°C

psychroduric

binary fission in bacteria

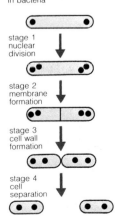

stage 1 nuclear division

stage 2 membrane formation

stage 3 cell wall formation

stage 4 cell separation

mesophilic bacteria bacteria (↑) which grow best at 30–40°C and which will not usually grow below 5°C or above 45–50°C.

mesophile (*n*) a mesophilic bacterium (↑).

psychrotrophs (*n*) psychrophiles (↑) and mesophiles (↑) as one group.

obligate aerobes bacteria (↑) which can only grow in a good supply of oxygen.

facultative aerobes bacteria (↑) which grow best in the presence of oxygen but can grow anaerobically (p.32).

obligate anaerobes bacteria (↑) which only grow in the absence of oxygen.

facultative anaerobes bacteria (↑) which grow best in the absence of oxygen but can grow aerobically (p.32).

binary fission the method of reproduction (p.107) of bacteria (↑). The nucleic acids (p.31) reproduce (p.107) themselves and the cell divides into two daughter cells (↓).

daughter cell a new cell produced by binary fission (↑).

graph showing growth rate of a microbial culture

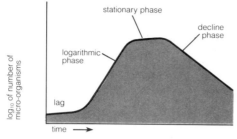

stationary phase

decline phase

logarithmic phase

lag

log₁₀ of number of micro-organisms

time ⟶

lag phase the period before reproduction (p.107) and increase in number of micro-organisms (p.147).

logarithmic phase the period of reproduction (p.107) and growth of micro-organisms (p.147) when the production of new organisms is logarithmic.

culture (*n*) a group or collection of bacteria (p.148) or other micro-organisms (p.147).

colony (*n*) a large group of growing micro-organisms (p.147).

colony count a measure of microbial (p.147) activity by growing a sample from contaminated (p.152) food on a medium such as agar (p.85) and counting the number of colonies (↑) of micro-organisms after a period of growth.

generation time the time between binary fissions (p.149) in micro-organisms (p.147). It may be as short as 12 minutes in some bacteria.

spore (*n*) a body produced by some micro-organisms (p.147) in conditions unfavourable to survival. Spores have a hard coat and can stand extreme conditions. Of the bacteria (p.148) only *Bacillus* and *Clostridium* form spores but these can survive even in cooked foods and lead to food spoilage (p.154) or food poisoning (p.152).

vegetative form a micro-organism (p.147) in cell form rather than spore (↑) form.

fungi (*n.pl.*) a group of plants having no chlorophyll (p.30). Many fungi are edible (p.111), e.g. truffles, mushrooms, and many are single-celled (p.147), e.g. moulds (↓) and yeasts (↓). **fungus** (*sing.*).

spore
formation in bacteria

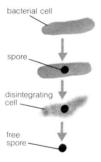

bacterial cell

spore

disintegrating cell

free spore

edible fungi mushrooms

mould (*n*) a group of multi-cellular (p.147) fungi (↑) which grow in thin thread-like strands called hyphae (↓). Each cell can grow alone and so moulds are classed as micro-organisms (p.147). Moulds can grow on foods and damage them but some are introduced into foods to give added flavour, e.g. in some cheeses (p.124). Most moulds reproduce (p.107) by forming spores (↑).

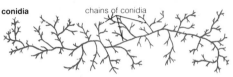

conidia   chains of conidia

conidia e.g. *Penicillium*

chains of conidia

hypha (*n*) a thread/strand of mould (↑). **hyphae** (*pl*).
sporangium (*n*) the hard case which protects a mould (↑) spore (↑).
conidium (*n*) an unprotected mould (↑) spore (↑). **conidia** (*pl*).
yeast (*n*) a single-celled (p.147) fungi (↑) which reproduces (p.107) by budding (↓). Many yeasts are used in fermented (p.34) foods such as bread and alcoholic beverages (p.93). Many yeasts, particularly those which grow as a bloom (p.60) on fruit, are able to grow in high acid foods and this can lead to spoilage of fruit and fruit juices (p.90), fruit preserves and meat (p.95).

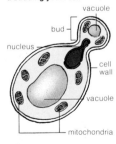

a budding yeast cell

vacuole
bud
nucleus
cell wall
vacuole
mitochondria

budding a process in yeast (↑) and some other organisms where a new individual is produced by part of the mother cell or individual growing out into a bud (p.105). The bud is eventually pinched off as a new organism.
osmophilic yeast a yeast (↑) which can grow in high concentrations (p.22) of sugars (p.44) or salts (p.16).
yeast extract a food product made from the water soluble (p.23) compounds of hydrolysed (p.13) yeast (↑) containing B vitamins and protein (p.61).
biological oxygen demand BOD. A measure of microbial (p.147) activity in a contaminated (p.152) material. Used mainly to measure the potential for contamination as other methods give more accurate determinations of microbial activity, e.g. colony counts (↑).

food poisoning any illness in man caused by eating foods containing harmful compounds. There are three main kinds of food poisoning; (1) chemical (↓), (2) biological (↓), and (3) microbiological food poisoning (↓).

chemical food poisoning food poisoning (↑) caused by harmful chemicals present in food, e.g. oxalic acid (↓) in rhubarb leaves, plant lectins (p.62). Some harmful chemicals in food may result from their use by man, e.g. compounds such as insecticides (↓) and herbicides (↓). Such compounds may be used to protect food crops during growth but, if still present at harvest (p.110), can cause a risk of food poisoning.

oxalic acid HOOCCOOH. A carboxylic acid (p.14) found in many plants in small quantities. It is contained in high concentrations in rhubarb leaves. Oxalic acid is a strong chelating agent (p.59) and can prevent absorption of metal ions (p.9) during digestion (p.41), e.g. in spinach. Also known as **ethanedioic acid.**

insecticide (n) a compound which kills insects. Insecticides are often sprayed on crops to reduce damage by insects.

herbicide (n) a compound which kills plants. Herbicides are used to kill plants which grow alongside food crops such as cereals (p.112).

weed-killer (n) = herbicide (↑).

biological food poisoning food poisoning (↑) caused by eating poisonous (p.41) natural foods, e.g. fungi (p.150).

microbiological food poisoning food poisoning (↑) caused by the products of micro-organisms (p.147) growing in foods. Food poisoning caused by contamination (↓) with bacteria (p.148) is the most common kind.

contaminated (adj) of foods which contain harmful substances or compounds which greatly reduce the quality of the food. **contaminant** (n).

causative agent a chemical, micro-organism (p.147) or other factor which is responsible for food poisoning (↑).

pathogenic (adj) of organisms or substances which cause disease in man.

biological food poisoning
poisonous fungi

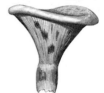

contamination chemical

blackening due to chemical reaction

canned goods

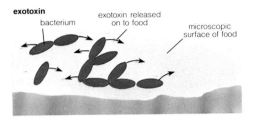

exotoxin

exotoxin released
on to food

bacterium

microscopic
surface of food

exotoxin (*n*) a toxin (p.41) released into food by
    bacteria (p.148) during growth and before the
    food is eaten.
endotoxin (*n*) a toxin (p.41) released by bacteria
    (p.148) eaten with contaminated (↑) food.
enterotoxin (*n*) a toxin (p.41) released by bacteria
    (p.148) in the alimentary canal (p.42).

enterotoxin

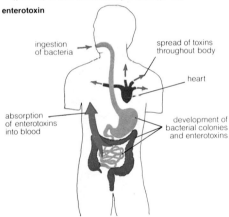

ingestion
of bacteria

spread of toxins
throughout body

heart

absorption
of enterotoxins
into blood

development of
bacterial colonies
and enterotoxins

heat-stable toxin a toxin (p.41) not damaged by
    heat.
infective (*adj*) of food-poisoning (↑) bacteria (p.148)
    which grow inside the body after being eaten.
mycotoxin (*n*) a toxin (p.41) produced by certain
    fungi (p.150).
botulism (*n*) a form of food poisoning (↑) caused
    by the toxins (p.41) of *Clostridium botulinum*.
    Botulism is rare but can cause death.

salmonella poisoning a form of food poisoning
(p.152) caused by the toxins (p.41) produced by
the *Salmonella* group of bacteria (p.148).
Salmonella poisoning is the most common form
of severe food poisoning in the Western world.

healthy carrier a person who carries a food
poisoning (p.152) organism in their gut (p.42)
without showing any of the symptoms (p.143) of
the illness usually caused by that organism. Such
people excrete the organism. Many animals can
be healthy carriers of food poisoning micro-
organisms, e.g. poultry (p.101), cattle, pigs, rats,
dogs and cats are all known to be occasional
carriers of *Salmonella*.

symptomless excretors = healthy carrier (↑).

convalescent carriers a person who excretes a
food poisoning (p.152) organism whilst
recovering from illness caused by that organism.
Excretion of the organism can continue for weeks
or even years.

incubation period the time between the eating of
a contaminated (p.152) food and the
appearance of the first symptoms (p.143). More
commonly used as the time for microbial (p.147)
growth.

cross-infection (*n*) the way in which a carrier of a
food poisoning (p.152) organism can spread it to
other individuals, e.g. the spread of *Salmonella*
infection between battery chickens.

food spoilage any damage caused to food which
does not necessarily make it harmful but makes it
inedible (p.111). There are three kinds of food
spoilage; enzymic browning (p.91), autolysis (↓)
and microbiological spoilage (↓).

autolysis (*n*) the breakdown of foods caused by
enzymes (p.68) within the food itself, e.g. the
unpleasant softening of over-ripe fruit.

microbiological spoilage any food spoilage (↑)
caused by the growth of micro-organisms (p.147)
on or in the food, e.g. the growth of mould
(p.151) on stored fruit.

microbial spoilage = microbiological spoilage (↑).

hygiene (*n*) the idea and practice of cleanliness.
Foods should always be handled in a hygienic
way. **hygienic** (*adj*).

food spoilage

mould growth on
surface of fruit

preserve (*v*) to extend the time that a food remains edible (p.111). **preservation** (*n*).

shelf-life (*n*) the expected length of time that a food product will maintain its quality and remain edible (p.111). Shelf-life depends on the method of preserving (↑) the food.

preservative (*n*) a compound added to foods to preserve (↑) them, e.g. sulphur dioxide.

deterioration (*n*) the process in which food quality is reduced due to spoilage (p.154), bad treatment or poor storage.

pasteurization (*n*) a process in which foods are heated to kill the vegetative forms (p.150) of bacteria (p.148). Temperatures of 65–100°C are used to pasteurize foods. Pasteurization can also be carried out with ionizing radiation (↓). **pasteurize** (*v*), **pasteurized** (*adj*).

sterilization (*n*) a process in which foods are treated to kill all forms of micro-organisms (p.147) and spores (p.150). Foods can be sterilized with high temperature treatment or with ionizing radiation (↓).

ionizing radiation high energy particles (p.24) given off by radioactive elements (p.8).

irradiation (*n*) ionizing radiation (↑), used for sterilization (↑) and pasteurization (↑).

ionizing radiation

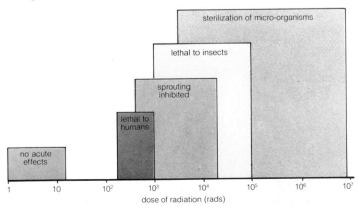

dose of radiation (rads)

canning

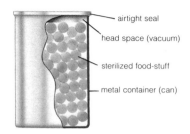

- airtight seal
- head space (vacuum)
- sterilized food-stuff
- metal container (can)

canning (*n*) a process in which food is sterilized (p.155) in cans (↓).

can (*n*) a metal container used for canning (↑) which can be sealed to prevent contamination (p.94) by micro-organisms (p.147).

laquer a material used to coat the inside surface of cans (↑) used in the canning (↑) process. Without lacquer coatings the metal of the can may be attacked by acids (p.14) and sulphur compounds leading to the production of black iron and tin sulphides which will discolour the food.

aseptic (*adj*) without contamination (p.94) by micro-organisms (p.147).

aseptic filling a process in UHT (↓) sterilization (p.155) in which the container of the food and the food product are sterilized separately and the containers are then filled aseptically (↑).

UHT ultra high temperature.

filling solution the liquid (p.22) used to fill cans (↑) near to the top before they are sealed in the canning (↑) process.

head space a small gap left at the top of cans (↑) before they are sealed in the canning (↑) process so that a partial vacuum (p.22) is formed in the space between the food and the top of the can.

retort (*n*) a piece of equipment used to heat foods for sterilization (p.155), pasteurization (p.155) or other thermal processing (p.87), e.g. spray drying (p.162). **retort** (*v*), **retortable** (*adj*).

hydrostatic retort a specialized retort (↑) of very large size in which a head of water is used to maintain pressure inside the retort allowing continuous throughput of cans (↑). Also known as **continuous retorts.**

bottling (*n*) a similar process to canning but using glass bottles instead of cans (↑), e.g. bottled milk.

HTST high-temperature-short-time. A process by which food is sterilized (p.155) at very high temperatures but for only very short times, e.g. 130°C for less than a minute. Pasteurization (p.155) can also be done by the HTST method, e.g. milk may be pasteurized at 72°C for 15 seconds.

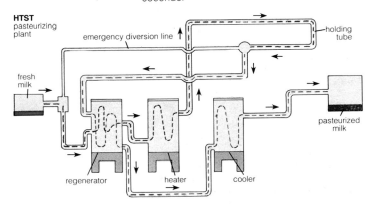

HTST pasteurizing plant

emergency diversion line ↑

holding tube

fresh milk →

pasteurized milk

regenerator   heater   cooler

decimal reduction time the time taken to reduce the number of micro-organisms (p.147) in a food sample to one tenth of the original amount when heating at a constant temperature.

D value = decimal reduction time (↑).

F value the sterilizing (p.155) value of a process. The F value is defined as the number of minutes at 121°C which will have a sterilizing effect equivalent to that of a sterilizing process.

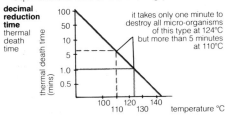

decimal reduction time
thermal death time

it takes only one minute to destroy all micro-organisms of this type at 124°C but more than 5 minutes at 110°C

thermal death time (mins)

temperature °C

refrigeration (*n*) the cooling of foods to low
temperatures to cause their preservation (p.155),
i.e. temperatures above 0°C but below 5°C. The
functions of most micro-organisms (p.147) are
considerably slower at this temperature and so
their growth is inhibited (p.70). **refrigerate** (*v*),
refrigerated (*adj*).

refrigerant (*n*) a chemical used in cooling devices
such as refrigerators (↑) and freezers (↓) which is
capable of carrying heat. Refrigerants are
generally able to change easily from gas (p.22)
to liquid (p.22) and back again. They are able to
absorb heat energy whilst changing state from
gas to liquid or vice versa and then to give up
that heat on changing back.

refrigerated sea water sea water chilled (↓) to
about −1°C and used to store fish. **RSW** (*abbr*).

chilling (*n*) the process by which food is preserved
(p.155) by being cooled to refrigeration (↑)
temperatures, i.e. 0–5°C.

cook-chill (*n*) the name given to a number of
similar processes in which food is cooked (p.87)
to sterilizing (p.155) or pasteurizing (p.155)
temperatures and then immediately chilled (↑)
and stored between 0°C and 3°C.

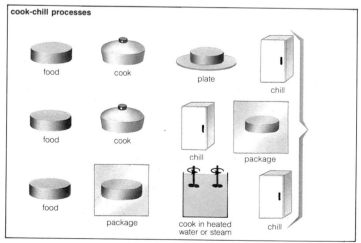

cook-chill processes

food cook plate chill

food cook chill package

food package cook in heated water or steam chill

freezing (*n*) a process in which food is preserved (p.155) by being cooled to temperatures below 0°C and normally as low as −20°C. The water and many of the components of foods solidify (p.22) during freezing. Because micro-organisms (p.147) require liquid (p.22) water for their life processes which normally function at higher temperatures, freezing inhibits (p.70) their growth. Foods such as plants are blanched (p.89) or cooked (p.87) before freezing but meats (p.195) and fish can be stored without treatment. **freeze** (*v*), **frozen** (*adj*).

immersion freezing a traditional method of freezing (↑) in which the food is lowered into a concentrated (p.22) salt (p.16) solution (p.22) containing enough ice to lower the temperature well below 0°C. It is very little used today.

plate freezing a widely used method of freezing (↑) foods. The food is packaged into cartons (p.94) or boxes which are placed between horizontal hollow refrigerated (↑) plates. Vertical plate freezers are also used for bulk products.

blast freezing a method of freezing (↑) foods in which cooled air (−40°C to −10°C) is blown around the food products by fans. Blast freezers are particularly useful for irregularly shaped foods which cannot be handled in a plate freezer (↑).

fluidized bed freezing a method of blast freezing (↑) in which small food products like beans and peas are suspended on and frozen by jets of cooled air. The process is often continuous, being carried out on a conveyer belt.

plate freezing

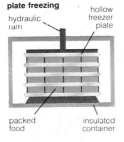

hydraulic ram
hollow freezer plate
packed food
insulated container

fluidized bed freezer

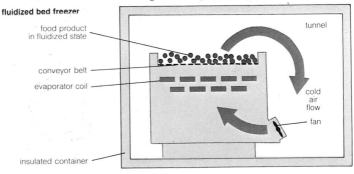

food product in fluidized state
conveyor belt
evaporator coil
insulated container
tunnel
cold air flow
fan

individually quick frozen of food products which have been rapidly frozen (p.159) as individual items, e.g. peas. Small food items can be quick frozen on trays or moving belts, as in fluidized bed freezing (p.159), prior to packing. This process is fast, and therefore leads to a better quality product, unlike freezing the products after packaging. **IQF** (*abbr*).

star marking a type of labelling (p.166) process in which the storage life of frozen (p.159) goods is indicated. One star represents storage at −6°C for up to one week, two stars storage at −12°C for up to one month and three stars storage at −18°C for up to three months.

rated freezer capacity the weight of food which can be frozen in a freezer in a 24 hour period. Rated freezer capacity is a convenient way of measuring the freezing power of an appliance.

ice crystals
freezer damage
slow freezing

cells

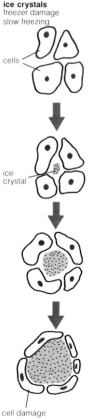

fast freezing

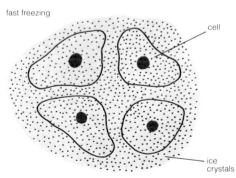

cell

ice
crystal

ice
crystals

cell damage

ice crystals crystals (p.11) of water formed in frozen (p.159) foods. Slow freezing (p.159) can result in the formation of very large ice crystals leading to structural damage of the food. Fast freezing leads to much smaller crytals and, therefore, less damage in the food.

super-cooled water non-crystalline water in frozen (p.159) materials. It usually results when samples are frozen very quickly, e.g. in liquid nitrogen.

freezer burn the drying of the surface of poorly wrapped frozen (p.159) foods due to sublimation (↓) of water.

sublimation (*n*) a process in which water or other liquids (p.22) change straight from the solid (p.22) phase (p.22) to the vapour (p.24) phase.

`**ice glaze** a film of ice produced over a food on freezing (p.159) to prevent freezer burn (↑). Commonly used with fish.

thaw (*v*) to melt (p.55) by returning frozen (p.159) materials to temperatures above 0°C. Thawing results in softening of the food and melting of ice crystals (↑).

cryogenic cooling freezing

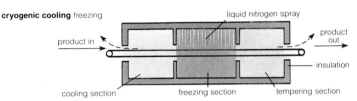

liquid nitrogen spray

product in

product out

insulation

cooling section   freezing section   tempering section

cryogenic cooling a process of cooling to chill (p.158) or freezing (p.159) temperatures using liquid (p.22) nitrogen.

cook-freeze (*v*) to cook (p.87), chill (p.158) and immediately freeze (p159) foods.

cold chain the chain of events which affect a frozen (p.159) or chilled (p.158) food during its life after production and before consumption. These events might include storage after production, refrigerated (p.158) transport, storage before purchase, storage before use. The cold chain is extremely important in controlling the quality of a frozen or chilled food. Temperature fluctuation (change) during the cold chain can lead to loss of quality in frozen or chilled foods due to thawing (↑) and re-freezing, increased oxidation (p.14) or growth of micro-organisms (p.147).

cold chain
typical temperature gradient

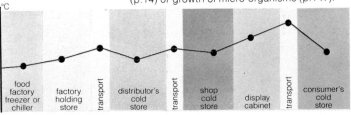

°C

| food factory freezer or chiller | factory holding store | transport | distributor's cold store | transport | shop cold store | display cabinet | transport | consumer's cold store |

dehydration using heated air

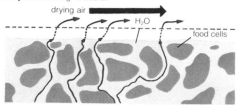

drying air ▮▮▮▮▮▮▮
H₂O
food cells

dehydration (*n*) a process of food preservation
(p.155) in which water is removed from the food
until only a small amount remains. There are
three main methods of dehydration: (1) methods
using heated air, e.g. allowing the food to dry
slowly in the sun, spray drying (↓), fluidized bed
drying (↓); (2) freeze drying (↓); (3) methods
using heated surfaces, e.g. roller drying (↓).

drying (*n*) = dehydration (↑).

freeze drying a process in which frozen (p.159)
food is placed in a vacuum (p.22) thereby
causing the rapid loss of water by sublimation
(p.161).

accelerated freeze drying freeze drying (↑) carried
out in conditions which speed up the process,
e.g. with heating.

lyophilizing (*n*) = freeze drying (↑). **lyophilize** (*v*),
lyophilized (*adj*).

spray drying a process by which liquids (p.22) are
dehydrated (↑) by being sprayed into a heated
container in which the droplets dry very quickly.

spray drying

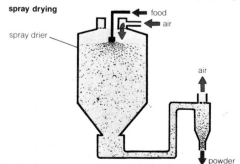

food
air
spray drier
air
powder

roller drying a process in which foods are dried (↑)
as pastes or particles (p.24) on heated rollers.
fluidized bed drying a process by which small
pieces of food are suspended and dried in warm
air.·

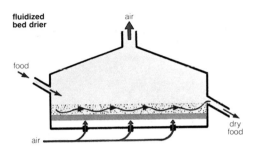

fluidized
bed drier

air

food

dry
food

air

regeneration (*n*) a process in which food is
returned to its original state for eating after chilling
(p.158), freezing (p.159) or dehydration (↑).
curing (*n*) a process by which meats and some
other foods are preserved (p.155) in salt (p.16)
solutions (p.22). The meat is steeped (↓) in a
brine (↓) solution to allow the salts to soak in and
then it is removed and stored.
steep (*v*) to soak.
brine (*n*) a solution (p.22) containing about 25%
sodium chloride (NaCl), 1% potassium nitrate
(KNO_3) and 0.1% sodium nitrite ($NaNO_2$) used for
curing (↑) meats (p.95) and in canning (p.156)
some vegetables.
nitrosamines (*n.pl.*) compounds produced from
nitrates (p.20) and nitrites in cured (↑) foods during
digestion (p.41); thought to be carcinogens (p.142).
dry curing a process of curing (↑) meat (p.95) with
solid salt (p.16) rather than brine (↑).
smoking (*n*) a process in which foods are preserved
(p.155) by hanging them in smoke and allowing
the surface to be coated with a layer of chemicals
from the smoke. This process is used today with
only a few foods and normally only to give
flavour rather than for the preservative effect.
pickling (*n*) a process by which foods (e.g.vegetables)
are preserved (p.155) in vinegar and brine (↑).

controlled atmosphere storage the storage of
foods, particularly fruit and vegetables, in
atmospheres of controlled gas (p.22) and water
vapour (p.24) content. The use of the correct
humidity (p.28) and different gases such as CO_2
can lead to the extention of the shelf-life (p.155)
of certain foods. CAS can be used either in large
storage rooms specially designed for keeping
foods or in packages which prevent the leakage
of the modified atmosphere. **CAS** (*abbr*).

controlled atmosphere packaging

heat seal of
transparent lid

modified atmosphere
inside pack

rigid tray

food
product

controlled atmosphere packaging the packaging
of foods in special containers in which the
atmosphere has been modified and can be
controlled. Containers used for CAP are often
made of special plastics (↓) like polythene (↓)
and polyamide (↓). *See* controlled atmosphere
storage (↑). **CAP** (*abbr*).

modified atmosphere packaging = controlled
atmosphere packaging (↑). **MAP** (*abbr*).

crytallized fruit a traditional method of preserving
(p.155) fruit particularly in the Middle East. Fruits
are cut into pieces which are coated in
concentrated (p.22) solutions (p.22) of sugar.
This causes dehydration (p.162) of the fruit
pieces through osmosis (p.28) of water out of the
plant tissue (p.32).

hypobaric storage a type of controlled atmosphere
storage (↑) in which fruit is stored under reduced
pressure (hypobaric). Ripening of the stored fruit
is delayed due to diffusion of ethene (p.110) (a
simple plant hormone (p.36)) into the
atmosphere.

PTFE a form of inert (p.12) plastic (↓) used widely in the production of wrapping and packaging material for foods.

plastic (*n*) a wide group of synthetic (p.13) chemical polymers (p.10) generally made from petroleum oil. Plastics are normally strong and can be moulded, stretched or rolled in a variety of shapes and forms and can be either rigid or flexible. Many foodstuffs are packaged in plastic containers ranging from bags and pouches (p.166) to cartons (p.94) and boxes. Common plastics used in packaging foods are polythene (↓) and polystyrene.

polythene (*n*) a plastic (↑), normally in the form of bags or pouches (p.166), used for packaging foods. Polythene can be either low density, which is strong, easily heat sealed (↓) and is resistant to water vapour (p.24) and temperatures up to 90°C or high density, which has the same properties as low density polythene but is resistant to temperatures up to and above 100°C. Polythene will not prevent the passage of oxygen, i.e. is not a good oxygen barrier (↓), and so does not inhibit (p.70) oxidative (p.14) deterioration (p.155) in foods.

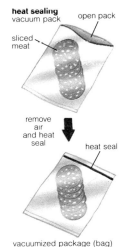

heat sealing vacuum pack   open pack

sliced meat

remove air and heat seal

heat seal

vacuumized package (bag)

heat sealing a method of sealing plastic (↑) containers by heating two adjoining layers or portions of the container until they melt together thereby forming a good seal.

oxygen barrier a material capable of preventing the passage of oxygen across itself, e.g. plastics (↑) such as polyamide (↓). Foods which need to be packed and stored under vacuum (p.22) must be sealed in containers made with oxygen barriers to prevent the leakage of oxygen into the container during storage of the food product.

polyamide (*n*) a plastic (↑) which can be made into films or layers and used in the production of packages which will resist the passage of oxygen. Polyamide is normally used as a laminate (↓) with other plastics like polythene and is particularly useful in the packaging of foods under vacuum (p.22).

nylon (*n*) = polyamide (↑).

laminated (*adj*) made up of several layers or films. **laminate** (*v*), **laminate** (*n*).

pouch (*n*) a flat, bag-like container made from plastic (p.165) or a laminate (p.165) of plastic and aluminium foil which is used to package sterilizable (p.155) foods. They are cheaper and more flexible than cans (p.156) but more difficult to seal and more easily damaged. Pouches can also be used for pasteurized (p.155) foods.

boil-in-the-bag a food product which is supplied par-cooked (partly cooked) or raw in a plastic (p.165) bag, normally made of high density polythene (p.165), which can be cooked by immersing the bag in boiling water. Products which can be treated in this way are limited as such high temperature cooking often damages such food as meats and fish.

label (*n*) a piece of paper attached to an object to give information about it. Food labels are required by law in many countries and can contain information such as the sell-by-date (↓), nutritive (p.137) content of the food and description and price of the product using a bar code (↓).

sell-by-date the latest date by which a preserved (p.155) or fresh food should be sold in order to maintain its quality. In many countries it is now required by law that the sell-by-date is shown on the package in which the foodstuff is sold, e.g. in Britain, the Food Labelling Regulations (1980) require chilled (p.158) foods, amongst other, to be clearly labelled with a sell-by-date.

best-before-date = sell-by-date (↑).

bar coding a system by which packaged food products can be labelled with a code made up of vertical bars. The bars can represent various important facts about the product, e.g. its description and price. Bar codes are read by moving them over an automatic scanner connected to a computer. This allows control of the sale and stock of food products and eliminates human error.

label

bar code

ingredients label

'best before' advice

nutritional advice

structure (*n*) (1) the arrangement of atoms (p.8), molecules (p.10), aggregates (p.25), cells (p.29), tissues (p.32) or parts of a substance or organism in three-dimensional space; (2) any object with a form or shape which can be described or defined.

function (*n*) the job done by a substance, cell (p.29), organelle (p.29), tissue (p.32) or organism, e.g. the function of an enzyme (p.68) is to catalyse (p.68) metabolic (p.144) reactions (p.12).

mechanism (*n*) the way in which a process (p.87) works or takes place, e.g. the mechanism of a chemical reaction (p.12).

stable (*adj*) of substances or organisms able to remain unchanged under different conditions.

unstable (*adj*) of substances or organisms which are destroyed or changed by certain processes (p.87), e.g. heating or irradiation (p.155).

unit (*n*) (1) a standard measurement, e.g. a gram, a metre, a degree centigrade; (2) a single object or structure which may make up part of the structure of some larger object, e.g. amino acids (p.61) are the units of a protein (p.61), monosaccharides (p.43) are the units of a polysaccharide (p.44).

sequence (*n*) (1) the order in which a set of events happen, e.g. reactions (p.12) in a metabolic pathway (p.144); (2) the order in which units (↑) are arranged in a line as part of a larger structure, e.g. the sequence of amino acids (p.61) in a protein (p.61).

property (*n*) a characteristic of a substance or object.

crude (*adj*) not pure.

test (*n*) a process (p.87) by which an object is examined in some way so that its properties (↑) are determined, e.g. chemical analysis (p.20), sensory analysis (p.131).

sample (*n*) a part of a larger object, area or group used in a test (↑) to measure, determine or describe the properties (↑) of the whole.

medium (*n*) a liquid (p.22) or solid (p.22) used by scientists to grow micro-organisms (p.147). Growth media contain all the necessary nutrients (p.137) for growth of the particular micro-organism, being studied. **media** (*pl*).

individual (*n*) a single organism, e.g. a bacterium (p.148), a plant, a person.

circumference (*n*) the distance around the outside of a circle.

diameter (*n*) the length of a line joining two sides of a circle but passing through its centre.

circumference and diameter

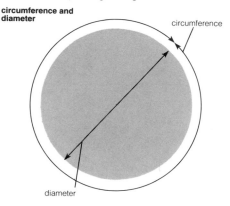

circumference

diameter

ratio (*n*) the product (↓) of one number divided by another.

product[2] (*n*) (1) a number resulting from a calculation; (2) a substance made during a chemical reaction (p.12); or (3) something made during a physical process (p.87), e.g. an object made in a factory or from a food raw material.

centrifugation (*n*) a process (p.87) by which liquid (p.22) samples are spun around at high speed to cause the accelerated settling of particles (p.24) in suspension (p.23).

oven (*n*) a heated container for cooking (p.87), e.g. roasting (p.87) and baking.

production line an arrangement in which the production of an object is carried out in stages, with each stage following the one before it in order.

optimum (*adj*) the best or most favourable. Usually used of the conditions for some process.

matrix a substance or place in which something takes form or is enclosed. Particularly used of connective tissues (p.97), the spaces between cells (p.29) or the inner structure of some foods.

International System of Units (SI)

PREFIXES

PREFIX	FACTOR	SIGN	PREFIX	FACTOR	SIGN
milli	$\times 10^{-3}$	m	kilo	$\times 10^{3}$	k
micro	$\times 10^{-6}$	μ	mega	$\times 10^{6}$	M
nano	$\times 10^{-9}$	n	giga	$\times 10^{9}$	G
pico	$\times 10^{-12}$	p	tera	$\times 10^{12}$	T

BASE UNITS

UNIT	SYMBOL	MEASUREMENT
metre	m	length
kilogram	kg	mass
second	s	time
ampere	A	electric current
kelvin	K	temperature
mole	mol	amount of substance
candela	cd	luminous intensity

COMMON DERIVED UNITS

UNIT	SYMBOL	MEASUREMENT
newton	N	force
joule	J	energy, work
hertz	Hz	frequency
pascal	Pa	pressure
coulomb	C	quantity of electric charge
volt	V	electrical potential
ohm	Ω	electrical resistance

Minerals

calcium a mineral, present in milk products, fish, hard water and in bread, which is required for healthy bones and teeth, to aid in the clotting of blood and in muscles. The average adult human requires 1.1 grams per day and the total body content is about 1000 grams.

phosphorus a mineral, present in most foods but especially cheese and yeast extract, which is required for healthy bones and teeth, and takes part in DNA, RNA and ATP metabolism. The average human adult requires 1.4 grams per day and the total body content is about 780 grams.

sulphur a mineral, present in foods containing proteins, such as peas, beans and milk products. It is required as a constituent of certain proteins, such as keratin and vitamins, such as thiamine. The average human adult requires 0.85 grams per day and the total body content is about 140 grams.

potassium a mineral present in a variety of foods, such as potatoes, mushrooms, meats and cauliflower, which is required for nerve transmission and acid-base balance. The average human requires 3.3 grams per day and the total body content is about 140 grams.

sodium a mineral present in a variety of 'salty' foods but especially table salt (sodium chloride), cheese and bacon, and which is required for nerve transmission and acid-base balance. The average human requires about 4.4 grams per day and the total body content is about 100 grams.

chlorine as chloride ions, a mineral found with sodium in table salt and in meats, which is required for acid-base balance and for osmoregulation. The average human adult requires about 5.2 grams per day and the total body content is about 95 grams.

magnesium a mineral present in most foods, but especially cheese and green vegetables, which is required to activate enzymes in metabolism. The average human adult requires about 0.34 grams per day and the total body content is about 19 grams.

iron a mineral present in liver, eggs, beef and some drinking water, which is an essential constituent of haemoglobin and catalase. The average human adult requires about 16 milligrams per day and the total body content is about 4.2 grams.

fluorine as fluoride, a mineral found in sea water and sea foods and sometimes added to drinking water. It is a constituent of bones and teeth and prevents tooth decay. The average human requires about 1.8 milligrams per day and the total body content is about 2.6 grams.

zinc a mineral found in most foods, but especially meat and beans, which is required as a constituent of many enzymes. It is also thought to promote healing. The average human adult requires 13 milligrams per day and the total body content is about 2.3 grams.

copper a mineral found in most foods, but especially in liver, peas and beans, which is required for the formation of certain enzymes. The average human adult requires about 3.5 milligrams per day and the total body content is about 0.07 grams.

iodine a mineral found in sea foods and some drinking water and vegetables, which is required as a constituent of thyroxine. The average human adult requires about 0.2 milligrams per day and the total body content is only about 0.01 grams.

manganese a mineral found in most foods, but especially tea and cereals, which is required in bones and to activate certain enzymes in amino acid metabolism. The average human adult requires about 3.7 milligrams per day and the total body content is only 0.01 grams.

chromium a mineral found in meat and cereals.

cobalt a mineral found in most foods, but especially meat and yeast products, which is an essential constituent of vitamin B_{12}. The average human adult requires about 0.3 milligrams per day and the total body content is as little as 0.001 grams.

APPENDIX THREE

Proteins

Recommended daily allowances of protein

AGE (YEARS)	WEIGHT (kg)	PROTEIN (g)
1–3	13	23
4–6	20	30
7–10	30	34
males		
11–14	44	45
15–18	61	56
19–22	67	56
23–51+	70	56
females		
11–14	44	46
15–18	54	46
19–22	58	44
23–51+	58	44
pregnancy		+30
lactation		+20

Vitamins

NAME	LETTER	MAIN SOURCES	FUNCTION	EFFECTS OF DEFICIENCY	FAT (F) OR WATER (W) SOLUBLE
retinol	A	liver, milk, vegetables containing yellow and orange pigments, e.g. carrots	light perception, healthy growth, resistance to disease	night blindness, poor growth, infection, drying and degeneration of the cornea	F
calciferol	D	fish liver, eggs, cheese, action of sunlight on the skin	absorption of calcium and phosphorus and their incorporation into bone	bone disorders, e.g. rickets	F
tocopherol	E	many plants, such as wheatgerm and green vegetables	cell respiration, conservation of other vitamins	in humans, no proved effect, may cause sterility, muscular dystrophy in rats	F
phylloquinone	K	green vegetables, egg yolk, liver	synthesis of blood clotting agents	haemorrhage, prolonged blood clotting times	F
thiamin	B_1	most meats and vegetables, especially cereals and yeast	coenzyme in energy metabolism	beri-beri, loss of appetite and weakness	W

riboflavine	B_2	milk, eggs, fish, green vegetables	coenzyme in energy metabolism	ulceration of the mouth, eyes and skin	W
niacin	B complex (B_2)	fish, meat, green vegetables, wheatgerm	coenzyme in energy metabolism	pellagra: skin infections, weakness, mental illness	W
pantothenic acid	B_5	most foods, especially yeast, eggs, cereals	coenzyme in energy metabolism	headache, tiredness, poor muscle co-ordination	W
pyridoxine	B_6	most foods, especially meat, cabbage, potatoes	release of energy, formation of amino acids	nausea, diarrhoea, weight loss	W
biotin	B complex (H)	most foods, especially milk, yeast, liver, egg yolk	coenzyme in energy metabolism	dermatitis	W
folic acid	B_c	green vegetables, liver, kidneys	similar to vitamin B_{12}	a form of anaemia	W
cobalamin	B_{12}	meats, e.g. liver, heart, herrings, yeast, some green plants	maturing red blood cells, growth, metabolism	a form of anaemia	W
ascorbic acid	C	citrus fruits, green vegetables	collagen formation and other metabolic functions	scurvy: tooth loss weakness susceptibility to disease, weight loss	W

Common food poisoning bacteria

NAME	SOURCE	SPORES	AEROBIC OR ANAEROBIC
Salmonella (many different types)	Human carriers, poultry, cattle, pigs, domestic pets, duck eggs, raw milk, etc.	No	aerobic
Staphylococcus aureus	Human carriers (especially nose and throat), animals, wounds, raw milk, etc.	No	faculatative aerobes
Clostridium perfringens	Human and animal intestines, soil (spores in dust and dirt), flies, etc.	Yes	anaerobic
Clostridium botulinum	Soil, marine muds, fish, some vegetables.	Yes	anaerobic
Bacillus cereus	Soil, dust, water, rice, cornflour, some other cereals.	Yes	aerobic

Some food additives
known or suspected to be harmful

NAME	EEC NUMBER[1]	REPORTED MEDICAL PROBLEMS
Tartrazine	E102	Urticaria[3], hayfever[3], asthma[3], hyperactivity
Quinoline yellow	E104	Hyperactivity
Yellow 2G	107[2]	Allergies[3] hyperactivity
Sunset yellow	E110	Allergies[3], hyperactivity
Cochineal	E120	Hyperactivity
Carmoisine	E122	Allergies, hyperactivity
Amaranth	E123	Allergies, hyperactivity
Ponceau red	E124	Allergies, hyperactivity
Erythrosine	E127	Phototoxicity
Red 2G	128[2]	Little known, hyperactivity
Patent blue V	E131	Allergies[3], hyperactivity
Indigo carmine	E132	Allergies, hyperactivity
Brilliant blue	133[2]	Hyperactivity
Caramel	E150	(Some forms may be suspect)
Black PN	E151	Little known, hyperactivity
Brown HT/FK	154–155[2]	Allergies[3], hyperactivity
Aluminium	E173	Linked with senile dementia
Benzoic acid family	E210–E218	Asthma, urticaria, hyperactivity
Sulphur dioxide	E220	Gut irritant
Sulphites	E221–E227	Dangerous to asthmatics, skin disorders
Biphenyls	E230–E232	May cause nausea in high doses
Hexamine	E239	Gastric upset, possible carcinogen
Nitrates } Nitrites	E249–E252	Converted to possible carcinogens in gut
Lactic acid	E270	Only problems in very young babies
Sodium proprionate	E281	Possibly causes migraine
Gallates	E310–E312	Gastric irritation, allergy[3], hyperactivity
Butylated anti-oxidants	E320–E321	Raises blood cholesterol/fat, hyperactivity
Carrageenan	E470	Reported to cause ulcerative colitis, products in gut may be carcinogenic
Gum arabic	E414	Very rare allergies
Mannitol	E421	Very rare hypersensitivity, nausea
Polyoxyethylene stearates	430–431[2]	Allergies in some susceptible people
Ammonium bicarbonate	503[2]	Gut irritant
Monosodium glutamate	621[2]	'Chinese restaurant syndrome'
Monopotassium glutamate	622[2]	'Chinese restaurant syndrome'
Potassium bromate	924[2]	Nausea in high doses
Chlorine	925[2]	Powerful irritant in high doses

[1] Food additive approved by the EEC commission and denoted by the prefix E.
[2] Not yet assigned an E prefix – awaiting EEC approval.
[3] Especially in people who are sensitive to aspirin or who already suffer from asthma.

Index

Edible plant material – non-fruit

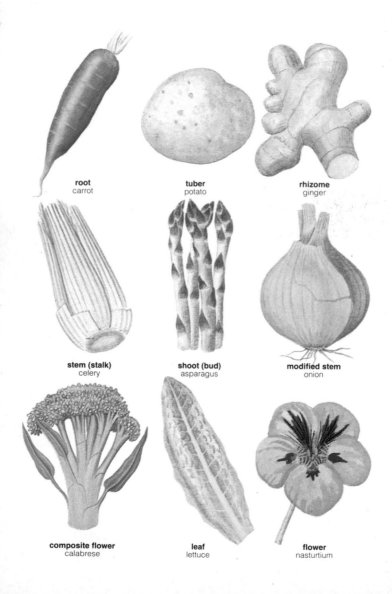

root
carrot

tuber
potato

rhizome
ginger

stem (stalk)
celery

shoot (bud)
asparagus

modified stem
onion

composite flower
calabrese

leaf
lettuce

flower
nasturtium